Administrative Regions until 1987

INTERNATIONAL BOUNDARY
ADMINISTRATIVE REGION BOUNDARY

Administrative Regions of Ethiopia until 1987.

Field guide to Ethiopian orchids

Field guide to Ethiopian orchids

Sebsebe Demissew   Phillip Cribb   Finn Rasmussen

First published in 2004 by
Royal Botanic Gardens, Kew,
Richmond, Surrey, TW9 3AB, UK

www.kew.org

Copy-edited by Beth Lucas.

Designed by Jeff Eden.

Typesetting and page layout by
Media Resources, Information Services Department,
Royal Botanic Gardens, Kew.

ISBN 1 84246 071 4

Printed by Compass Press Ltd

For information or to purchase other Kew titles please visit
www.kewbooks.com or email publishing@kew.org

Contents

Preface

It is a pleasure to be asked to write a foreword to *Field Guide to Ethiopian Orchids* by Sebsebe Demissew, Phillip Cribb and Finn Rasmussen. This is a fitting recognition of the strong collaboration that has developed over two decades, between botanists working in the National Herbarium of Addis Ababa University, Ethiopia, and those in Uppsala in Sweden, the Royal Botanic Gardens at Kew, England, and the Botanical Museum of Copenhagen University, Denmark.

In 1969, I returned from the University of North Wales, Bangor, having completed a doctorate in plant ecology full of enthusiasm to start research into the rich and diverse ecosystems of my country. My advisor, Prof. Greig-Smith, returned with me so that we could discuss how we could collaborate. But very soon I found a major barrier to ecological research – I could only identify a few of the plants I was dealing with because Ethiopia lacked any modern Flora. The *Flora of Tropical East Africa* was of some help, but most of the families, and particularly many of the challenging families for an ecologist, were not yet published. Other botanists in Europe were also aware of the lack of a modern Flora for Ethiopia, which then also included Eritrea, although it had been one of the first countries to be botanically explored. Following a mini-symposium in the 1970 meeting of AETFAT (Association pour l'étude taxonomique de la flore d'Afrique tropicale), two committees under the leadership of Prof. Olov Hedberg of Uppsala University were formed and a project proposal for such a Flora was prepared. This came to Addis Ababa University, where I had become Dean of the Science Faculty, and I had the pleasure of helping modify the proposal to accommodate Ethiopian interests, and obtain the approval of the Ethiopian government. One of my main concerns was to ensure that the writing of a Flora of Ethiopia would be accompanied by the building up of capacity in both manpower and facilities in the National Herbarium. Thanks to the long and consistent financial support from the Swedish International Development Cooperation Agency (Sida), all these concerns have been met.

The publication of complete accounts for all the monocotyledon families has been one of the major achievements of the Flora Project. These appeared in 1995 and 1996 in Volume 6 (all monocotyledon families, excluding grasses) and Volume 7 (grasses) of the *Flora of Ethiopia and Eritrea.* An important result of the publication of the Flora volumes has been the stimulation of further research and collecting. This is well reflected in this book. The account of the Orchidaceae in Volume 6 covered 154 species, while the present book has 167 species described and illustrated. However, as the authors point out in their introduction, the orchids of Ethiopia are still poorly known with many of the species represented by ten or fewer collections.

This field guide is also welcome because orchids are plants that attract the attention of specialists, including unscrupulous collectors. It is, therefore, important to have a well-illustrated guide in order to keep a better-informed check on the status of these plants.

One fascinating feature of the orchids of Ethiopia is that over eighty percent are terrestrial plants, and many of these occur in wetlands. These are areas often regarded as 'waste' or 'unused' land that should be developed. Wetlands contribute greatly to keeping the hydrological balance in an area, and the presence of orchids can help support arguments for better conservation of such areas for the benefit of these plants and the many other wetland species associated with them.

I much look forward to having a copy of this book in my hand when I get the opportunity to go into the field and remember the challenges I faced more than thirty years ago in identifying the plants around me. This is a very valuable extension of the work of the Flora project from which my younger colleagues can continue to benefit for many years to come.

Tewolde Berhan Gebre Ezgiabher
General Manager
Environmental Protection Authority
Addis Ababa, Ethiopia

April 22 2003

Acknowledgements

We would like to thank the editors of the *Flora of Ethiopia and Eritrea*, Prof. Inga Hedberg and Ms Sue Edwards for giving the permission to use the information on Volume 6 of the *Flora of Ethiopia and Eritrea*. The Curator and Keeper of the National Herbarium in Addis Ababa (ETH), the Keeper of the Herbarium, Royal Botanic Gardens, Kew (K) and their staff have kindly supported this work and allowed access to their collections for which we are most grateful.

The fine line drawings are the work of Juliet Beentje, Susanna Stuart-Smith, Maureen Church and Stella Ross-Craig; Anne Farrer drew the large maps, Justin Moat the distribution maps; and the following provided photographs: Ib Friis, Mike Gilbert, Christoff Herrmann, David Menzies, Börge Pettersson, Mats Thulin, Tesfaye Awas, Isobyl and Eric la Croix, Joyce Stewart, L.B. Segerbäck, B. Bytebier and Lucy and John Tanner. We would like to thank them all for their generosity.

We also extend our appreciation to the following colleagues and friends who provided comment and helpful advice in the preparation of this work: Simon Owens, Sally Bidgood, Sarah Thomas, Suzy Dickerson and John Harris.

The support for field work in Ethiopia over a number of years by the Norwegian NUFU and the Swedish Sida-SAREC is highly appreciated and acknowledged.

Funding for an orchid conservation workshop in Addis Ababa in September 2003 from the Zürich Foundation for Orchid Conservation inspired this work. We are particularly grateful to Dr. Roman Kaiser and Dr. Albert Pfeiffer for their support in gaining this funding.

Sebsebe Demissew
Phillip Cribb
Finn Rasmussen

2004

Introduction

Orchidaceae is the fifth largest family in the *Flora of Ethiopia and Eritrea*, with an estimated 167 species in 37 genera (Table 1). Three of the five subfamilies of orchids, namely Orchidoideae, Vanilloideae and Epidendroideae, are represented in the flora.

Orchid numbers compare rather poorly with those of the much richer floras of tropical East Africa (679 species), West Africa (413 spp.), Central Africa (517 spp.) and South Africa (425 spp.). The region's orchid flora shows a distinct cline of decreasing richness from the south-west to the north-east, mirroring the similarly decreasing rainfall patterns. The majority of Ethiopian orchids are terrestrials of the grasslands and woodlands. Only 16% are epiphytes (growing on trees) or lithophytes (growing on rocks), and these are mostly confined to the wetter and more forested areas of the west and south-west of the country.

A number of botanists and collectors, such as Richard Quartin-Dillon, Georg Wilhelm Schimper and Georg Schweinfurth, visited Ethiopia in the 19th century and collected orchids there. These represent some of the earliest collections of tropical African orchids. Despite this and their diversity and popularity, the orchids of the region are sparsely represented in herbaria. The majority of the remaining species are known from ten or fewer collections. Some orchids admittedly are difficult to collect e.g. *Nervilia*, which flowers after the leaf has withered, and the fleshy and leathery leafed species which make poor herbarium specimens. Others flower at times when collecting is difficult such as during the rainy season. However, we can state that the orchids of the region are generally poorly collected and little understood. The forests, particularly of south-west Ethiopia, are under-explored and the orchids of the Gamo Gofa and Illubabor regions are very poorly represented in herbaria. New records are likely as the remoter regions become better explored. No orchids have yet been recorded from the low, arid lands of the Afar region.

The larger genera

Only three orchid genera are represented by more than ten species in Ethiopia: the terrestrial genera *Habenaria* (47 species) and *Eulophia* (25 species), and the predominantly epiphytic *Polystachya* (12 species).

A cosmopolitan genus, *Habenaria* is particularly well-represented in tropical Africa with possibly 300 species. With twice as many species as *Eulophia,* the next largest genus in the region, it boasts almost one-third of the Ethiopian orchid flora. Several sections are represented by only one or two species in Ethiopia. For example, *Habenaria cornuta*, of section *Ceratopetalae*, is a terrestrial found in *Acacia-Commiphora* woodland and *Habenaria macrura*, of section *Macrurae*, is a species of seasonally wet grassland. For these and many others, Ethiopia represents the northern extent of the range of the species. In contrast, Ethiopia appears to be a centre of diversity for section *Multipartitae*, with 12 species found here. Some of the most impressive members of the genus belong in this section such as the appropriately named *H. excelsa*, *H. praestans*, and *H. decorata*.

Eulophia is represented by 25 species. Most of these, such as *E. cucullata* and *E. odontoglossa*, are widespread species in tropical Africa, while a few, such as *E. streptopetala* and *E. speciosa*, are found throughout East Africa south to South Africa. The remarkable *E. petersii*, also widespread in East and South Africa, grows in some of the most arid conditions tolerated by orchids. It somewhat resembles a *Sanseveria* when not in flower and is equally tolerant of drought because of its stout pseudobulb and succulent leaves.

Polystachya, another pantropical genus with its centre of diversity in tropical Africa, is mainly epiphytic but some species can also be found growing as lithophytes. Twelve species are found in Ethiopia. Most are woodland or forest plants such as *P. paniculata* but some such as *P. eurychila* and *P. steudneri* can survive relative aridity by losing their leaves in the dry season. Several of the Ethiopian species are widespread in tropical Africa, such as *P. cultriformis* and *P. bennettiana*, but most are more restricted in their distribution, a frequent feature of *Polystachya* in Africa.

Many of the other genera, such as *Platycoryne, Brachycorythis, Disa, Satyrium, Angraecum, Microcoelia* and *Tridactyle*, are common in tropical Africa but only represented by one or two species in Ethiopia.

Biogeographical affinities

Three-quarters of Ethiopian species also occur in the *Flora of Tropical East Africa* (FTEA) region of Kenya, Uganda and Tanzania. A comparison of the two regions shows that whilst the FTEA orchid flora has a high proportion of epiphytes, the Ethiopian flora is dominated by terrestrials.

Seventy-two species are common to both Ethiopia and West Africa. When pan-African species are excluded, the West African influence on the flora is less marked with only 17 species being found in West Africa and Ethiopia.

Twenty of the 23 orchid species recorded from the Arabian Peninsula also occur in Ethiopia (Cribb, 1979; Robbins, 1992), the majority of them being confined to the highlands of Yemen. Amongst these are the remarkable *Epipactis veratrifolia*, often found growing in seepages in near desert-like conditions. An orchid of the Middle East and eastern Mediterranean, it is the only species from that region found in Africa. It is also found in Somalia whose other ten species are also all found in Ethiopia (Pettersson, 1995).

Of particular interest has been the recent discovery of *Vanilla roscheri* by Sally Bidgood and Ib Friis. Collected from Gamo Gofa at 1050 m, it was growing in sandy soil in *Acacia* bushland with scattered evergreen shrubs and an annual rainfall of c. 900 mm, similar to its habitats in Kenya and coastal Tanzania (Bidgood & Cribb, 2001). In this area it was apparently locally common. Apart from suggesting that a thorough survey of the orchid flora would reveal much that is currently unrecorded, the disjunct distribution of this species between the East African coast and southern Ethiopia is noteworthy and reminiscent of the distribution of some other flowering plants such as *Portulaca petersii* (Portulacaceae), *Ruellia amabilis* (Acanthaceae) and *Abelmoschus ficulneus* (Malvaceae).

Endemism

Tournay (1972) reported 50 endemics in the Ethiopian orchid flora of 123 species, i.e. about 40% endemism. The current study indicates that this is an overestimate: 28 species, some 15% of the region's orchids, are endemic. Of these, 22 are terrestrial and only 6 are epiphytic. If those orchids shared by Ethiopia and Yemen are included another five species can be added. A further ten Ethiopian species are endemic to the region if adjacent Kenya, Uganda and Sudan are included.

Nine species of *Habenaria* are endemic, including *Habenaria vollesenii*, *H. gilbertii*, *H. montolivaea*, *H.*

taeniodema and *H. excelsa*, all montane grassland species. Amongst the other endemic terrestrial species are three of *Disperis*, two species each of *Roeperocharis* and *Eulophia*, *Satyrium aethiopicum*, *Disa facula*, *Holothrix unifolia* and *Liparis abyssinica* (known only from the type collection).

Endemic epiphytes include *Stolzia grandiflora*, a large-flowered member of a tiny, creeping, peperomia-like genus related to the Asiatic genus *Eria*; three species of *Polystachya*, including the common *P. caduca*, probably the most frequently collected Ethiopian orchid; the poorly understood *Cyrtorchis erythraeae*; and the charming but rare *Diaphananthe candida* which resembles a small *Aerangis*.

Conservation Although Ethiopian orchids have been relatively well documented over many years compared with other families (Richard 1848–1851; Tournay, 1972; Cribb & Thomas, 1997), they are still rather poorly understood. For example, little can be said of their biology, ecology and conservation status, although many appear to be rare. We suspect the epiphytes are particularly threatened because of widespread felling of Ethiopia's remaining woodlands and forest. However, the evidence is flimsy and more detailed surveys are urgently needed to assess the status of these species.

We expect that novelties will continue to be added to the Flora and suggest that specialised field work aimed at the orchids would be particularly productive.

Orchids have been widely used elsewhere to highlight the importance of plant conservation and to protect orchid-rich habitats, thereby protecting other plants and animals in those places. A better understanding of the orchids of the region may be useful in helping protect its biodiversity (Hagsater & Dumont, 1996). Orchids, particularly if they are showy, rare or endemic, can be used as an education tool, to raise public support and funding for conservation, to pin-point areas worthy of conservation, and to encourage eco-tourism.

Note: We have provided conservation status information based on herbarium data and our field observations in Ethiopia. The criteria used for IUCN Red Listing have been used as far as possible. However, the IUCN Red List categories have not been used, these generally being applied to species throughout their ranges. Flowering times given are those from Ethiopia, except where no authenticated material has been seen. In which case, flowering times from adjacent Sudan or Uganda have been added.

Table 1. Numbers of Ethiopian orchids by genus

Genus	Current no. of species	Terrestrial (T) or Epiphytic (E)	Endemics
Aerangis	5	E	
Ancistrorhynchus	1	E	
Angraecopsis	2	E	
Angraecum	3	T/E	
Bolusiella	1	E	
Bonatea	2	T	
Brachycorythis	3	T	
Bulbophyllum	4	E	
Calyptrochilum	1	E	
Cheirostylis	1	T	
Corymborkis	1	T	
Cynorkis	2	T	
Cyrtorchis	2	E	1
Diaphananthe	6	E	1
Disa	7	T	1
Disperis	6	T	3
Epipactis	2	T	
Eulophia	25	T	2
Graphorkis	1	E	
Habenaria	47	T/E	11
Holothrix	7	T	1
Liparis	3	T/E	1
Malaxis	1	T/E	
Microcoelia	1	E	
Nervilia	3	T	
Oberonia	1	E	
Oeceoclades	2	T/E	
Platycoryne	1	T	
Platylepis	1	T	
Polystachya	12	T/E	3
Pteroglossaspis	1	T	
Rangaeris	1	E	
Roeperocharis	3	T	2
Satyrium	7	T	1
Stolzia	2	E	1
Tridactyle	2	E	
Vanilla	1	T	
Total	**167**		**28**

Table 2. Endemic and near endemic orchids in the Ethiopian Flora

Species	Distribution	Altitudinal range (m.)
Angraecopsis holochila Summerh.	GD, SD, TU, Uganda	1500–2300
Cyrtorchis erythraeae (Rolfe) Schltr.	EW, KF, SD	1350–1700
Diaphananthe adoxa F.N.Rasm.	BA, IL, KF, SD, WG, Kenya, Uganda	1300–2300
Diaphananthe candida P.J.Cribb	KF, SD, WG	2000–2100
Diaphananthe schimperiana (A.Rich.) Summerh.	AR, BA, GD, HA, KF, SD, SU, Sudan, Uganda	2100–2850
Diaphananthe tenuicalcar Summerh.	GD, GJ, KF, SD, SU, WG, Kenya, Uganda	1350–2400
Disa facula P.J.Cribb et al.	WG	1400–1500
Disa pulchella Hochst. ex A.Rich.	AR, BA, GD, GJ, SD, Yemen	1800–3800
Disperis crassicaulis Rchb.f.	GD, HA, SD, SU, TU	2000–2500
Disperis galerita Rchb.f.	GD, GJ, SD	2000–3800
Disperis meirax Rchb.f.	GD	3500–3800
Eulophia abyssinica Rchb.f.	EW, GD, GJ, KF, SD, TU	2250–2600
Eulophia albobrunnea Kraenzl.	AR, BA, GJ, KF, HA, SD, SU, WU	1600–2500
Habenaria aethiopica S.Thomas & P.J.Cribb	GJ, KF, SU, WG	2250–2450
Habenaria antennifera A.Rich.	GJ, KF, SU, TU, Yemen	2000–3300
Habenaria cavatibrachia Summerh.	BA, Kenya, Uganda	2100–2700
Habenaria cultiriformis Kraenzl.	GG, HA, KF, SD, TU, Yemen	1140–2200
Habenaria cultrata A.Rich.	EW, SU, TU, Yemen, Oman	1700–2100
Habenaria decorata A.Rich.	AR, GD, GJ, SU, WU, Uganda, Kenya	2200–3800
Habenaria decumbens S.Thomas & P.J.Cribb	AR, SD	1900–2600
Habenaria excelsa S.Thomas & P.J.Cribb	GJ	3150–3500
Habenaria gilbertii S.Thomas & P.J.Cribb	SU	2100–2300
Habenaria macrantha A.Rich.	AR, BA, EW, GD, GJ, SD, SU, WU, Uganda, Kenya, Somalia, Yemen	1900–3100

continued >>

Habenaria montolivaea Kraenzl. ex Engl.	AR, GD, SU, WU	1000–2600
Habenaria perbella Rchb.f.	EW, GD, TU	1200–1500
Habenaria platyanthera Rchb.f.	GD	unknown
Habenaria quartiniana A.Rich.	BA, GD, SU, TU, Uganda, Kenya	2100–2600
Habenaria rivae Kraenzl.	SD	unknown
Habenaria taeniodema Summerh.	GJ	3150–3500
Habenaria tricruris (A.Rich.) Rchb.f.	AR, EW, GJ, SD, SU, TU	2000–3000
Habenaria vollesenii S.Thomas & P.J.Cribb	SD	1200–1575
Holothrix squammata (A.Rich.) Rchb.f.	AR, GD, GJ, SU, SD, Uganda, Sudan	2400–2800
Holothrix unifolia (Rchb.f.) Rchb.f.	GD, SU	2500–2900
Liparis abyssinica A.Rich.	TU	unknown
Polystachya aethiopica P.J.Cribb	AR, SU	1350–2200
Polystachya caduca Rchb.f.	AR, BA, GD, KF, SD, SU, WG	2400–2600
Polystachya eurychila Summerh.	AR, KF, SD, SU, Uganda, Kenya	1700–2000
Polystachya rivae Schweinf.	EW, KF, SU, WG	1770–2490
Roeperocharis alcicornis Kraenzl.	AR, GD, TU	c. 2600
Roeperocharis urbaniana Kraenzl.	GD	c. 2750
Satyrium aethiopicum Summerh.	KF, SD, SU, TU, WG	2000–2500
Satyrium brachypetalum A.Rich.	AR, SD, SU, TU, Yemen	2000–2500
Stolzia grandiflora P.J.Cribb	BA, SD, SU, WG	1900–2850

AR = Arsi
BA = Bale
EE = Eritrea East
EW = Eritrea West
GD = Gonder
GG = Gamo Gofa
GJ = Gojam
KF = Kefa
SD = Sidamo
SU = Shewa
TU = Tigray
WG = Wellega
WU = Welo

References

Bidgood S. & Cribb, P.J. (1999). Vanilla Orchidaceae newly reported from Ethiopia. *Kew Bull.* 52 (4): 378.

Cribb, P.J. (1979). Orchids of Arabia. *Kew Bull.* 33 (4): 651–678.

Cribb, P.J. & Thomas, S. (1997). Orchidaceae. In S. Edwards, Sebsebe Demissew & I. Hedberg, eds. *Flora of Ethiopia & Eritrea* 6: 193–307.

Hagsater, E. & Dumont, V., eds. (1996). *Orchids. Status Survey and Conservation Action Plan.* IUCN Species Survival Commission, Gland, Switzerland & Cambridge, UK.

Pettersson, B. (1995). Orchidaceae. In M. Thulin, ed. *Flora of Somalia* 4: 70–76.

Richard, A. (1848–1851). *Orchidaceae. Tentamen Florae Aethiopiae* 2: 281–303.

Thomas, S. (1992). Additional records of Orchidaceae for the Flora of Arabia. *Kew Bull.* 47(4): 721–724.

Tournay, R. (1972). Orchidaceae. In G. Cufodontis, *Enumeratio Plantarum Aethiopiae Spermatophyta* 2: 1597–1622.

Geography

Located between 3° and 15° N and 33° and 48° E, Ethiopia is a large and diverse country. It is a country of dramatic contrasts, with the Semien and Bale Mountains reaching over 4000 m and the Danikil Desert sinking below sea level. In some parts the highlands are dissected and form sharp peaks on the horizon, giving the impression, as one observer remarked, of "a country with a table upside-down".

Although much of the interior consists of highland plateaux, these are interrupted by deep gorges and valleys formed by large rivers such as the Abay (Blue Nile), Tekeze, Mereb, Awash, Omo, Genale, Wabi-Shebelle and Baro. The great Rift Valley divides the country in two: the western and northern highlands on one side and the south-eastern highlands on the other. Both these highland systems gradually decrease in elevation to form vast arid or semi-arid lowlands in the east, west, and southern parts of the country. The Rift Valley is itself a dramatic sight with land falling away into a broad gorge up to 2000 m deep. Whereas the edge of the Rift Valley is temperate grassland, the bottom of the gorge is a land of hot, dry, savannahs and deserts, interspersed with large lakes.

The natural landscape of Ethiopia has been greatly affected by the activities of man, especially his agricultural activities over many centuries.

Geology

Ethiopia's geology is based on an old crystalline block, originally part of an immense area that stretched from India to Brazil and formed part of the ancient super-continent Gondwanaland. The hard crystalline rocks are mainly granites and gneisses, and contain many valuable mineral deposits.

Precambrian rocks with ages of over 600 million years, found in parts of Tigray, Gonder, Gojam, Harerge, Sidamo, Bale, Illubabor and Wellega, are the oldest rocks in the country and form the basement on which younger formations lie (Ethiopian Mapping Authority, 1988). These include a wide variety of sedimentary, volcanic and intrusive rocks that have been metamorphosed to varying degrees. In the southern and western parts of the country, where they are predominantly granites and gneisses, they have been more strongly metamorphosed than their counterparts in the north. This is mainly due to the fact that the rocks in the north have been subjected to relatively low temperatures.

Towards the end of the Precambrian, a major uplift followed by a long period of erosion took place. Sediments deposited during the Palaeozoic interval (which lasted some 375 million years) had been largely removed by erosion.

During the Mesozoic (starting about 225 million years ago), subsidence occurred and the sea initially spread over the Ogaden then gradually extended further north and west. As the depth of the water increased, sandstone, mudstone and limestone were deposited. As the land mass was uplifted, sedimentation ended in the western parts of the country with the deposition of clay, silt, sand, and conglomerate from the land as the sea receded. In the south-eastern parts, gypsum and anhydrite deposits were precipitated.

In the Early Cenozoic (which began 65 million years ago), extensive faulting took place. However, the major displacement along the fault systems of the Red Sea, Gulf of Aden and East African Rifts occurred later during the Tertiary. Faulting was accompanied by widespread volcanic activity which led to the deposition of vast quantities of basalt, especially over the western half of the country. The great table-lands of the centre, north-west and east were built up by the spreading deep basalts. Their erosion has also produced the spectacular mountain scenery of the Semien Mountains with its dramatic jagged pinnacles, precipices and gorges. The faulting was followed by, and alternated with, the eruption of large amounts of ash and coarser material, forming the Trap Series.

Ethiopia's largest lake, Lake Tana, is the result of damming a large natural drainage area on the western plateau. The Blue Nile disgorges from the Lake and has carved a deep, steep-sided gorge as it runs in an arc around the eastern and southern sides of the Lake. The Tis Isat Falls close to the lake are where the Blue Nile runs over the edge of an ancient lava flow.

More recent volcanic activity is associated with the development of the Rift Valley, activity being concentrated within the Rift and along the edges of the adjoining plateaux and also in the Danikil Depression. Plugs of old volcanoes are seen throughout the highlands. Hot springs are frequent in many areas and earth tremors are not uncommon. The recent sediments (conglomerate, sand, clay, and reef limestone) which accumulated in the Afar Depression and at the northern end of the main Rift Valley are of Quaternary age.

Climate

Ethiopia exhibits a wide variation in its climate. In the Semien and Bale Mountains snowfalls are periodically experienced; in the Danikil Desert, daytime temperatures can reach 50°C or more. As a rule, the central highlands have a temperate climate without extremes of temperature. In contrast, the eastern and southern lowlands are hot and dry, the western lowlands are moist and hot and the southern rift valley has a hot and seasonally moist climate. Elevation, temperature and rainfall are thus the major influences on Ethiopia's climate.

Temperature In Ethiopia, temperature is mainly influenced by latitude and elevation (Ethiopian Mapping Authority, 1988). Ethiopia lies between 3°N and 18°N, within the tropics. True tropical temperatures are encountered in low-lying areas, especially in the east, south and west along the country's borders. Much of the central part of Ethiopia is mountainous, with the highest peaks reaching to over 4000 m. Because the highlands mostly lie over 1500 m altitude, typical tropical temperatures are unusual in many places. Thus temperature decreases towards the interior, with the mean annual temperature ranging from about 40°C in the lowlands to less than 10°C in the highest areas. Extremes of temperature can be experienced, depending on one's location.

Rainfall The rainfall pattern in Ethiopia is influenced by two rain-bearing wind systems, one bringing the Monsoonal (westerly) winds from the South Atlantic and the Indian Ocean and the other bringing easterly winds from the Arabian Sea. The two systems alternate, producing different rainfall regimes in different parts of the country.

Four major rainfall regimes are experienced in Ethiopia:

A The central, eastern and north-eastern areas of the country receive a bimodal (two peaks) rainfall pattern. The small spring rains between February and May come from the Arabian Sea and the big summer rains,

Major rainfall patterns in Ethiopia

A. bimodal rainfall: February-May/June-September
B. unimodal rainfall
C. bimodal rainfall: September-November/March-May
D. scanty rainfall but November-February rainfall maximum

between June and September, mainly come from the South Atlantic, but also from the Indian Ocean. Both these periods of rain decrease in length northwards.

B The south-western and western areas of the country have a single wet period, usually between February and April. This is influenced by both the wind systems coming from the South Atlantic and the Indian Ocean. The length of the wet season decreases northwards. The south-western highlands receive the highest mean annual rainfall of over 2700 mm. Most of the native orchid species are found here.

C The southern and south-eastern parts of the country receive a distinctly bimodal rainfall with the first peak between September and November and the second between March and May. There is a distinct dry spell between the two peaks. This rainfall pattern is mainly influenced by the monsoon winds from the Indian Ocean.

D The northern part of the rift valley and adjacent areas have scanty rainfall in a diffused pattern, but with significant amounts of winter rainfall between November and February. These areas receive a mean annual rainfall of 200 mm or less. No orchid species have so far been found here.

References

Ethiopian Mapping Authority (1988). *National Atlas of Ethiopia*. Berhanena Selam Printing Press, Addis Ababa.

Tesfaye Haile (1986). *Climatic variability and surface feedback mechanisms in relation to Saheleo-Ethiopian Droughts*. MSc Dissertation. Dept of Meteorology, University of Reading, UK.

Workineh Degefu (1987). Some Aspects of Meteorological Drought in Ethiopia. In: M. Glantz (ed.), *Drought and Hunger in Ethiopia*. pp. 23–36.

Vegetation

The vegetation of Ethiopia is divided into eight major types (Dove, 1890; Pichi-Sermolli, 1955, 1957; Breitenbach, 1963; White, 1983; Friis, 1992; Sebsebe Demissew et al., 1996; CSE, 1997). Friis & Sebsebe Demissew (2001) provided a detailed account of how these vegetation types have been conceptualized by various earlier authors and the various attempts at making vegetation maps of Ethiopia based on these types. The vegetation map shown overleaf is that currently used in Ethiopia, but a better and more complete map based on more field observations is urgently needed.

The map shows eight vegetation types, one more than listed in the following paragraphs (the riverine and marshy vegetation is not shown as it is not possible to distinguish it graphically). In Sebsebe Demissew et al. (1996) and CSE (1997), the vegetation type indicated as evergreen scrub in the map and description was depicted as a distinct zone surrounding the dry evergreen montane forest and grassland. However, it has since been realized that the evergreen scrub is not found in a distinct zone of its own, but rather in a mosaic of vegetation types, mainly associated with the dry evergreen montane forest or as a derived vegetation where moist evergreen montane forest has been destroyed. Thus, evergreen scrub is considered here to belong mostly in subtype 1 of the dry evergreen montane forest and grassland complex. The connection between evergreen bushland and moist evergreen montane forest was suggested by Friis et al. (1982) in their description of the vegetation of SW Ethiopia.

Orchids occur in most vegetation types, except the dry lowland deserts and semi-deserts and the higher afro-alpine vegetation, where no orchid species have yet been found. Orchids are commonest in moist evergreen montane forest and in swamp vegetation.

1 **Desert and semi-desert scrubland.** This vegetation is highly drought tolerant, but due to external influences, such as human and animal trampling around watering points, the land can locally be completely devoid of vegetation. The soils are often alluvial, associated with

Vegetation Types

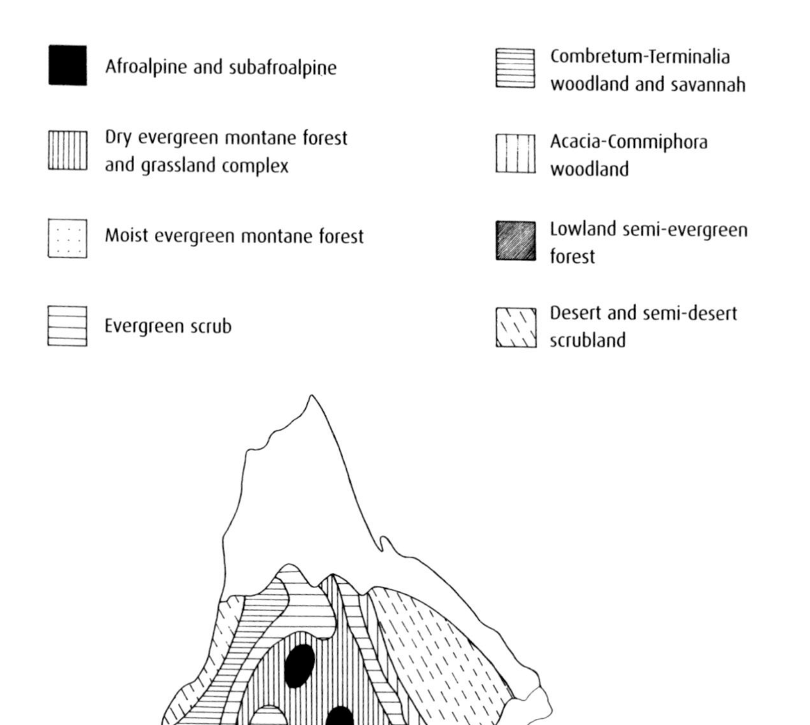

Map showing vegetation types of Ethiopia after Sebsebe Demissew *et al.* (1996) and CSE (1997)

Desert scrubland. Bokol Mayo-Dolo, Sidamo

the Awash River basin, but may also be derived from basaltic rocks and lava flows. This vegetation is found below an altitude of 500 m. Small trees and shrubs occur, including species of *Acacia* (Leguminosae), *Commiphora* (Burseraceae), *Boscia*, *Cadaba*, *Maerua* (all Capparaceae) and *Zizyphus* (Rhamnaceae), as well as succulents, including species of *Euphorbia* (Euphorbiaceae) and *Aloe* (Aloaceae). Most grass species (Poaceae) are annual, such as *Dactyloctenium aegyptium*, although another characteristic species, *Panicum turgidum*, is perennial. This vegetation type occurs in the Afar Depression, the Ogaden, around Lake Chew Bahir and the Omo delta.

No orchids are known from this type of habitat.

2 **Acacia-Commiphora woodland.** This vegetation is characterised by drought-resistant trees and shrubs, either deciduous or with small, evergreen leaves. The trees and shrubs form an almost complete stratum and include species of *Acacia*, *Commipora*, *Balanites*

Acacia-Commiphora woodland. Sof Omar, Bale

(Balanitaceae), *Capparis* (Capparaceae), *Combretum* and *Terminalia* (both Combretaceae). But due to the small leaves in the tree and shrub layers, light penetrates to the ground, which often has a rich flora. Thus, the ground cover is rich in subshrubs, including species of *Acalypha* (Euphorbiaceae), *Barleria* (Acanthaceae), *Aerva* (Amaranthaceae); geophytes, including a number of species of *Chlorophytum* (Anthericaceae); and succulents, including a number of *Aloe* species. Sometimes no trees are present, and in such cases the vegetation should be referred to as *Acacia-Commiphora* bushland. This vegetation type occurs in the northern, eastern, central and southern part of the country between 900 and 1900 m.

A number of terrestrial orchids with rhizomatous underground parts and thick roots commonly occur here, including *Eulophia petersii*.

3 **Moist evergreen montane rainforest.** This vegetation type is in most cases characterised by one or more closed strata of evergreen trees that may reach a height of 30 to 40 m. Sometimes only the lower stratum remains, due to the removal of the canopy. This vegetation type can be further divided:

Subtype 1 is what is traditionally referred to as ***Afro-montane rainforest.*** According to Friis (1992) these forests occur in the south-western part of the Ethiopian Highlands between 1500 and 2600 m, with an annual rainfall between 700 and 1500 mm. The Harenna Forest on the southern slopes of the Bale Mountains is the easternmost example of this forest. The northernmost examples are patches of this type of forest in Wellega. The canopies typically contain a mixture of *Podocarpus falcatus* (Podocarpaceae) and broad-leaved species, but *P. falcatus* is never a single dominant and becomes gradually more infrequent towards the south-west in Kefa and Illubabor as the rainfall increases, while *Pouteria (Aningeria) adolfi-friederici* (Sapotaceae) becomes more prominent in the same direction. The other very important conifer in Ethiopia, *Juniperus procera* (Cupressaceae) does not occur here. The drier parts of these forests are floristically very similar to those in the central parts of the Ethiopian highlands, while the south-western, humid parts approach subtype 2 (see below). The more or less continuous canopy below *Podocarpus falcatus* and *Pouteria (Aningeria) adolfi-friederici* consists of medium-sized trees,

10–30 m tall, including species of *Albizia* (Leguminosae), *Ilex mitis* (Aquifoliaceae), *Prunus africana* (Rosaceae), *Ocotea kenyensis* (Lauraceae), *Polyscias fulva* (Araliaceae), *Sapium ellipticum* (Euphorbiaceae), *Syzygium guineense* subsp. *afromontanum* (Myrtaceae) and *Olea capensis* subsp. *hochstetteri* or *O. welwitschii* (Oleaceae). *Pouteria (Aningeria) adolfi-friederici* or *Podocarpus falcatus* are the only emergent species from the 20–30 m high canopy. The smaller trees and the large shrubs form a discontinuous stratum. The moist humid forests have dense stands of tree ferns, *Cyathea* (Cyatheaceae) in the ravines. Epiphytes include ferns, orchids, and *Peperomia* (Piperaceae). The bamboo *Arundinaria alpina* (Poaceae) is not uncommon at higher altitudes in this area.

A number of epiphytic orchids commonly occur here. Examples include: *Angraecum*, *Angraecopsis*, *Bulbophyllum*, *Liparis deistelii*, *Polystachya bennetttiana* and *Stolzia repens*.

Afro-montane forest. Bonga, Kefa

Subtype 2 is ***transitional rainforest.*** Scattered examples of these forests are known from the western escarpment of the Ethiopian Highlands in Wellega, Illubabor and Kefa, where the humidity from the rain-bearing southwesterly winds is highest. The forests occur between 500 and 1500 m elevation. The rainfall is close to 2000 mm per year or higher (up to 2700 mm), with rain all the year round. The transitional rainforests are most similar in physiognomy and composition to the humid broad-leaved Afro-montane rainforests of south-western Ethiopia (described as subtype 1 above), with the addition of species from the lowland forest (described overleaf).

The canopy includes *Pouteria (Aningeria) altissima*, *Anthocleista schweinfurthii* (Loganiaceae), *Ficus mucuso* (Moraceae), and species of *Garcinia* (Clusiaceae/ Gutiferae), *Manilkara* (Sapotaceae) and *Trilepisium* (Moraceae).

A number of terrestrial and epiphytic orchids also occur here, such as *Bulbophyllum* and *Polystachya* species.

Transitional forest. Sheko, Kefa

4 **Lowland semi-evergreen forest.** This vegetation type is similar to the dry peripheral semi-deciduous Guineo-Congolian forest of Friis (1992). Such forests are restricted to the Baro Lowlands of western Illubabor, and have only been recognised in fairly recent years. They occur between 450 and 650 m on sandy soils, but usually with a high groundwater table, with average maximum temperatures of 35–38°C, mean annual temperatures of 18–20°C, and annual rainfall between 1300 and 1800 mm. They are semi-deciduous, with a 15–20 m tall, more or less continuous canopy in which *Baphia abyssinica* (Leguminosae) (endemic to southwestern Ethiopia and adjacent areas of the Sudan) is dominant, mixed with less

Lowland semi-evergreen forest. Gog, Gambela

common species including *Celtis toka* (Ulmaceae), *Diospyros abyssinica* (Ebenaceae), *Malacantha alnifolia* (Sapotaceae), *Zanha golungensis* (Sapindaceae) and species of *Lecaniodiscus* (Sapindaceae), *Trichilia* (Meliaceae), and *Zanthoxylum* (Rutaceae). Species of *Alstonia* (Apocynaceae), *Antiaris* (Moraceae), *Celtis*, and *Milicia* (Moraceae) emerge from the main canopy. Below the closed canopy is a more or less continuous stratum of small trees and shrubs. The ground is mostly covered by thick litter, and there are apparently very few subshrubs or herbaceous species on the forest floor, for example the widespread forest grass *Streptogyna crinita* (Poaceae).

A number of terrestrial and epiphytic orchids occur here.

5 **Combretum-Terminalia woodland and savannah.** This vegetation type is characterised by small to moderate-sized trees with fairly large deciduous leaves. These include *Boswellia papyrifera* (Burseraceae), *Anogeissus leiocarpus* (Combretaceae) and *Stereospermum kunthianum* (Bignoniaceae) and species of *Terminalia*, *Combretum* and *Lannea* (Anacardiaceae). The solid-stemmed lowland bamboo *Oxytenanthera abyssinica* (Poaceae) is prominent in river valleys (and locally on the escarpment) of western Ethiopia. The ground cover is a tall stratum of perennial grasses (Poaceae), including species of *Cymbopogon*, *Hyparrhenia*, *Echinochloa*, *Sorghum* and *Pennisetum*. In western Ethiopia where the grass biomass is largest, this vegetation type has been burned annually for such a long time that the plants show clear adaptations to fire, and it must be assumed not to be adversely affected by controlled annual fires. This vegetation type occurs along the western escarpment of the Ethiopian Plateau, from the border region between

***Combretum-Terminalia* woodland. Bewa Mountains, Benishangul-Gumuz**

Ethiopia and Eritrea to western Kefa and the Omo Zone. It is the dominant vegetation in what is now Benshangul-Gumuz and Gambella, and the Dedessa valley in Wellega in Oromya, where it occurs at 500–1900 m. At the upper limit it frequently abuts on to afro-montane moist evergreen forest. It penetrates into the Ethiopian plateau along the large river valleys. *Combretum-Terminalia* woodland also occurs as a comparatively narrow zone in Central and Eastern Ethiopia between the *Acacia-Commiphora* woodland and bushland and the vegetation on the plateau. However, in these parts of Ethiopia this vegetation type usually has too little extension to be shown on a map.

A number of terrestrial orchids with rhizomatous underground parts and fusiform roots commonly occur here, for example, *Eulophia guineensis* and *E. streptopetala*.

6 **Dry evergreen montane forest and grassland complex.** This vegetation type represents a complex system of successions involving extensive grasslands rich in legumes, shrubs and small to large-sized trees to closed forest with a canopy of several strata. Four distinct facies have been recognized (Friis, 1992; Friis & Sebsebe Demissew, 2001).

Subtype 1. Undifferentiated Afro-montane forest. The forests on the plateaux can be seen as a gradient from wet to dry types, and generally the vegetation on the plateaux appears now as a mosaic. The mosaic consists of humid sites, where areas of forest and evergreen (or semi-evergreen) bushland are now largely replaced by derived vegetation due to agriculture, and more well-drained sites or sites in rain shadow, where the original vegetation had been wooded grassland, woodland or deciduous bushland (see subtype 3). The undifferentiated Afro-montane forests are either *Juniperus-Podocarpus* forests, or tend towards single dominant *Podocarpus* or *Juniperus* forests, both with an element of broad-leaved species. They occur especially on the plateaux of Shewa, Welo, Sidamo, Bale and Harerge at altitudes between 1500 and 2700 m, with annual rainfall between 700 and 1100 mm. Presently, the few larger patches still extant appear widely separated by areas of cultivation and wooded grassland. The canopy is usually dominated by *Podocarpus falcatus* with *Juniperus procera* as co-dominant, *Croton macrostachyus* (Euphorbiaceae), *Ficus* spp. and *Olea europaea* subsp. *cuspidata*. There is usually a rather well-developed stratum of small to medium-sized trees. Scrambling species and true lianas are common. Epiphytes include species of *Peperomia*, ferns and orchids. The ground cover is rich in ferns, grasses, sedges, and small herbaceous dicotyledons. At the upper limit of this type of forest, between 3000 and 3400 m, there is often a more open type of woodland or evergreen bushland with *Erica arborea* (Ericaceae), *Gnidia glauca* (Thymelaeaceae), *Hagenia abyssinica* (Rosaceae), *Hypericum revolutum* (Guttiferae), *Jasminum stans* (Oleaceae), *Myrica salicifolia* (Myricaceae), *Myrsine africana* (Myrsinaceae), *Myrsine (Rapanea) melanophlöeos* (Myrsinaceae), *Rosa abyssinica* (Rosaceae), and *Nuxia congesta* (Loganiaceae). If classified according to physiognomy, this would be referred to as evergreen bushland. Clumps of the mountain bamboo, *Arundinaria alpina* may occur, but on the plateaux the species does not appear to form extensive stands such as are found in East Africa. *Albizia schimperiana* and *Acacia abyssinica* may form very dense stands, a dense *Albizia-Acacia* woodland or a forest, the canopy of which may be mixed with broad-leaved species. Lianas and a ground cover of partly hygrophilous herbs may occur. These subtypes are frequently seen as features of forest margins or seral stages in forest regeneration.

A number of orchids occur in this zone, such as epiphytes *Polystachya caduca*, *P. bennettiana* and *Stolzia repens* and terrestrial species including *Habenaria decora* and *H. quartiniana*.

Dry evergreen montane forest. Wefwasha - Ankober, Shewa

Dry Afro-montane forest. Slope of Lake Ashenge, Tigray

Subtype 2. Dry single-dominant Afro-montane forest of the Ethiopian Highlands. According to Friis, this forest occurs especially on the plateaux in Tigre, Gonder, Welo and Harerge at altitudes between (1600–) 2200 and 3200 (–3300) m with an annual rainfall between 500 and 1500 mm. The typical dominant species in the upper storey of these forests is *Juniperus procera*, with *Olea europaea* subsp. *cuspidata* and a number of other species below. Sometimes the juniper trees can be rather scattered and the forest is characteristic of *Juniperus* woodland with discontinuous evergreen undergrowth.

A few orchids, including *Polystachya bennettiana*, occur in this zone.

Subtype 3. Afro-montane woodland, wooded grassland and grassland. This includes the natural woodlands and wooded grasslands of the plateau with *Acacia abyssinica*, *A. negrii*, *A. pilispina*, *A. bavazanoi* and *A. montigena*,

Afro-montane woodland. Ambasel Mountains, Welo

some of which are endemic to the Ethiopian highlands. This vegetation type occurs either on well drained sites or on areas with black cotton soil which may be flooded during the rains, and must be assumed to have formed a mosaic with the forests and evergreen bushland vegetation of the plateau before the influence of man.

A few orchids occur in this zone.

Subtype 4. Dry single-dominant Afro-montane forest of the Eastern escarpments and transition between single-dominant Afro-montane forest and East African evergreen and semi-evergreen bushland. This is a transitional vegetation which includes a range of physiognomic types, from typical forest to evergreen scrub with dispersed trees, but floristically the whole range is connected. The forest-like types exist on the eastern escarpment of the Ethiopian Highlands, and on the northern escarpment of the Somali Highlands; the bushland-like types with trees occur scattered in the previous areas, on the south-eastern slopes of the eastern Highlands extending along the mountain chain into northern Somalia. Throughout its range it occurs on rocky ground with unimpeded drainage at altitudes between 1500 m and 2400 m and with an annual rainfall of between 400 and 700 mm. The dry juniper forests of Sidamo occur at altitudes between 1500 and 2000 (–2200) m with an average rainfall between 400 and 700 mm. In an open stratum of smaller trees, species of the following genera occur: *Acokanthera* (Apocynaceae), *Apodytes* (Icacinaceae), *Barbeya* (Barbeyaceae), *Berchemia* (Rhamnaceae), *Brucea* (Simaroubaceae), *Cadia* (Leguminosae), *Halleria* (Scrophulariaceae), *Mimusops* (Sapotaceae), *Olea*

Subafroalpine vegetation. Semien Mountains, Gonder

europaea subsp. *cuspidata*, *Pistacia*, *Rhus* (both Anacardiaceae), *Schrebera* (Oleaceae), *Sideroxylon* (Sapotaceae) and *Tarchonanthus* (Asteraceae). The dry juniper forests of Sidamo, rich in endemic species, form the southern extension of this type. Other characteristic species include (in Harerge) *Dracaena ellenbeckiana* (Dracaenaceae), *Buxus hildebrandtii* (Buxaceae) and *Barbeya oleoides*. In the north, *Dracaeana ombet* may occur at the lower edge of this forest type.

A few orchids occur in this zone.

7 **Afro-alpine and sub-Afro-alpine vegetation.** This vegetation type is characterised by small trees, shrubs and shrubby herbs at the lower altitudes and giant herbs, small herbs and grasses (the five distinctive Afro-alpine life-forms of Hedberg (1951)). The evergreen shrubs include *Erica arborea* and *Hypericum revolutum*. Typical perennial herbs include a number of species of *Helichrysum* (Asteraceae). *Lobelia rhynchopetalum* (Lobeliaceae) is alone among the giant rosette herbs with stems. The grasses mainly belong to tribes dominant in the temperate regions, and include species of *Festuca, Poa* and *Agrostis*. This vegetation type occurs in areas above 3200 m. These occupy the highest mountains in the country: the Choke Mts in Gojam, the Semien Mts in Gonder/Tigray, the high mountains in Tigray and Welo (in the north), the Guge Highlands in Gamo Gofa (in the south-west), the Chilalo mountain in Arsi and the Bale mountains in Bale (in the south-east).

A few terrestrial orchids, including *Disa pulchella*, *Habenaria lefebureana* and *Satyrium schimperi* have been recorded in this zone.

Afro-alpine vegetation. Senaiti Plateau, Bale Mountains National Park

8 **Riparian and swamp vegetation.** This vegetation type consists of at least two physiognomically different types: riverine and riparian forest with faster moving water, and open, almost treeless swamp vegetation with stagnant or slowly moving water almost throughout the year. Typical trees in riverine forest include *Celtis africana, Ficus sycomorus* and *Mimusops kummel.* The large areas of grassland on the plateau, which are only inundated in the rainy season, are here classified as montane grassland and wooded grassland with vegetation type 6, subtype 3.

The riverine and riparian forest vegetation of the study area is very variable, and the floristic composition is dependent on altitude and geographical location. Common trees in these forests are species of *Ficus, Celtia africana, Lepisanthes senegalensis* (Sapindaceae), *Salix* (Salicaceae), *Trichilia emetica, Diospyros mespiliformis, Mimusops kummel, Syzygium guineense, Tamarindus indica* (Leguminosae), *Acacia albida, Tamarix nilotica* (Tamaricaceae), *Breonadia salicifolia* (Rubiaceae), and *Phoenix reclinata* (Arecaceae). There is often a shrub layer, and lianas and vascular epiphytes occur. The ground cover includes grasses, ferns, and a few herbaceous dicotyledons.

The riparian vegetation at larger lakes may include *Acacia albida*, species of *Ficus, Phoenix reclinata* and *Aeschynomene elaphroxylon* (Leguminosae).

The swamps are dominated by sedges, grasses and many herbs. Orchids are frequent in swampy areas, especially *Eulophia* species such as *E. caricifolia* and *E. angolensis*, and some *Habenaria, Disa* and *Satyrium* species.

Riparian vegetation. Close to Abay River from Asosa side, Benishangul-Gumuz

Swamp vegetation. Benishangul-Gumuz

References

Breitenbach, F. von. (1963). *The Indigenous Trees of Ethiopia*. 2nd revised and enlarged edition. 306 pp. Ethiopian Forestry Association, Addis Ababa.

CSE, (1997). *The Conservation Strategy of Ethiopia (CSE). The Resource Base, its utilization and planning for sustainability*. Federal Democratic Republic of Ethiopia, Environmental Protection Authority, Addis Ababa.

Dove, K. (1890). Kulturzonen von Nord-Abessinien. Peterm. *Geo. Mitteil*. Ergänzungsband 21 (no. 97).

Friis, I. (1992). Forests and forest trees of northeast tropical Africa — their natural habitats and distribution patterns in Ethiopia, Djibouti and Somalia. *Kew Bull.*, Additional Series, No. 15 (pp. i–iv & 1–396). HMSO, London.

Friis, E., Rasmussen, F. N. and Vollesen, K. (1982). Studies in the Flora and vegetation of southwest Ethiopia. *Opera Bot*. 63: 1–70.

Friis, I. & Sebsebe Demissew (2001). Vegetation maps of Ethiopia and Eritrea. A review of existing maps and the need for a new map for the Flora of Ethiopia and Eritrea. In: I. Friis & O. Ryding (eds.), *Biodiversity*

Research in the Horn of Africa Region, Proceedings of the 3rd International Symposium on the Flora of Ethiopia and Eritrea. Biol. Skrif. 54: 399–439.

Hedberg, O. (1951). Vegetation belts on the East African mountains. *Svensk Bot. Tidsskr.* 45: 140–202.

Pichi-Sermolli, R.E.G. (1955). Tropical East Africa (Ethiopia, Somalia, Kenya, Tanganyika). Chapter VI, *Plant Ecology. Reviews of Research.* pp. 302–360 in UNESCO Arid Zone Research.

—— (1957). Una carta geobotanica dell'Africa Orientale (Eritrea, Ethiopia, Somalia). *Webbia* 13: 15–132 & 1 map.

Sebsebe Demissew, Mengistu Wondafrash & Yilma Dellellegn (1996). Ethiopia's natural resource base. pp. 36–53 in Edwards, S. (ed), *Important Bird Areas of Ethiopia. A First Inventory.* 300 pp. Ethiopian Wildlife and Natural History Society, Addis Ababa.

White, F. (1983). The vegetation of Africa. A descriptive memoir to accompany the UNESCO/AETFAT/UNSO vegetation map of Africa. With 4 coloured maps (1:5 000 000). *Natural Resources Research* 20: 1–356.

Countryside near Addis Ababa

ORCHIDACEAE

Perennial, terrestrial, mycotrophic, epiphytic or lithophytic *herbs* or rarely scrambling *climbers*, with rhizomes, root-stem tubers or rootstocks with mycorrhizal fungi in the roots and often elsewhere. Growth either sympodial or less commonly monopodial. *Stems* usually leafy, but leaves often reduced to bract-like scales, one or more internodes at the base often swollen to form a "pseudobulb"; aerial, assimilating adventitious roots, often bearing one or more layers of dead cells called a velamen, are borne in epiphytic species. *Leaves* glabrous or occasionally hairy, entire except at the apex in some cases, alternate or occasionally opposite, often distichous, frequently fleshy and often terete or canaliculate, almost always with a basal sheath which frequently sheaths the stem, sometimes articulated at the base of the lamina and sometimes with a false petiole. *Inflorescences* erect to pendent, spicate, racemose or paniculate, one- to many-flowered, basal, lateral or terminal, the flowers rarely secund or distichously arranged. *Flowers* small to large, often quite showy, hermaphrodite (or rarely monoecious and polymorphic outside Africa), sessile or variously pedicellate, most often twisted through 180 degrees, occasionally not twisted or twisted through 360 degrees. *Ovary* inferior, unilocular and the placentation parietal, (or rarely trilocular and the placentation axile but not in Africa). *Perianth* epigynous, of two whorls of three segments. The outer whorl of sepals usually free but sometimes variously adnate, the median (dorsal) often dissimilar to the laterals, the laterals sometimes adnate to the column foot to form a saccate, conical or spur-like mentum. The inner whorl comprising two lateral petals and a median lip; petals free or rarely partly adnate to sepals, similar to sepals or not, often showy; lip entire, variously lobed or two- or three-partite, ornamented or not with calli, ridges, hair cushions or crests, with or without a basal spur or nectary, margins entire to laciniate. *Column* short to long, formed from stylar and filamentous tissue, with or without a basal foot, occasionally winged or with lobes or arms at apex or ventrally; anther one (or rarely two or three in extra-African taxa), terminal or ventral on column, cap-like

Eulophia streptopetala

or opening by longitudinal slits; pollen in tetrads, agglutinated into discrete masses called pollinia; 2, 4, 6 or 8 pollinia, mealy, waxy or horny, sectile or not, sessile or attached by caudicles, a stipe or stipites to one or two sticky viscidia; stigma three-lobed, the midlobe often modified to form a rostellum, the other lobes either sunken on the ventral surface of the column behind the anther or with two lobes porrect. *Fruit* a capsule, opening laterally by three or six slits; seeds numerous, dust-like, lacking endosperm, sometimes markedly winged.

The classification of the family is currently the subject of some debate, particularly the number of subfamilies that should be recognised and the placement in those of certain tribes, subtribes and genera. The classification of Pridgeon *et al.* (1999, 2001, 2003), which is strongly supported by recent molecular, embryological and morphological analyses, is followed here. They recognise five subfamilies: Apostasioideae, Cypripedioideae, Vanilloideae, Orchidoideae and Epidendroideae. Only the last three families are found in Africa, all being present in Ethiopia. This classification differs slightly from that followed in the *Flora of Ethiopia and Eritrea* account (Cribb & Thomas, 1997) where Spiranthoideae, now included in Orchidoideae, was recognised as a distinct subfamily, and *Vanilla* was included in subfamily Epidendroideae.

165 species in 37 genera are reported from Ethiopia and Eritrea. Twenty-eight species, including six epiphytes, are endemic.

References

Cribb, P.J., Thomas, S. (1997). Orchidaceae. In S. Edwards, T. Mesfin & I. Hedberg, eds. *Flora of Ethiopia & Eritrea* 6: 193–307.

Pridgeon, A.M., Cribb, P.J., Chase, M.A.& Rasmussen, F.N., eds.(1999, 2001, 2003) *Genera Orchidacearum* I–III. Oxford Univ. Press.

ARTIFICIAL KEY TO THE GENERA OF ETHIOPIAN ORCHIDACEAE

1.	Plants leafless	2
–	Plants with leaves; rarely flowering when leafless	3
2.	Plants twig epiphytes; stem very reduced; roots clustered, photosynthetic	**29. Microcoelia**
–	Plants scrambling lianas; stem elongate, photosynthetic; roots solitary from the nodes, non-photosynthetic	**1. Vanilla**

3. Plants terrestrial 4
- Plants epiphytic or lithophytic 24

4. Anther attached at its base to column; plants with tuberous roots 5
- Anther attached by its apex to column; plants with a fleshy rhizome, pseudobulbs or elongate stem 14

5. Flowers with 2 spurs on hood **10. Satyrium**
- Flowers with a single spur on hood or lacking a spur on hood 6

6. Dorsal sepal with an erect or pendent spur **9. Disa**
- Dorsal sepal lacking a spur; lip or lateral sepals sometimes spurred 7

7. Lip with a basal saccate to cylindrical or clavate spur; lateral sepals not spurred or pouched 8
- Lip lacking a spur or saccate base, enclosed within hood, much reduced but often with complex appendages; lateral sepals with saccate or spur-like pouches **11. Disperis**

8. Flowers yellow or orange **7. Platycoryne**
- Flowers not so coloured 9

9. Flowers pink or purple, often glandular **4. Cynorkis**
- Flowers white or green, rarely flushed with violet 10

10. Stigmas 2-lobed; petals erect, wing-like either side of dorsal sepal, often with an undulate margin **8. Roeperocharis**
- Stigmas not 2-lobed; petals simple or 2-lobed 11

11. Stigma lobes sessile 12
- Stigma lobes elongate, clavate 13

12. Lip more or less fused to column at base **2. Holothrix**
- Lip free from the column **3. Brachycorythis**

13. Lateral sepals and front lobe of petals fused to the stigma lobes **6. Bonatea**
- Lateral sepals free from stigma lobes **5. Habenaria**

14. Plants with creeping short to elongate, fleshy or woody rhizomes; pollinia mealy or much divided into massulae; anther opening lengthwise 15
- Plants with pseudobulbs, stems or underground corm-like rhizomes or tubers 18

15. Stems woody, bamboo-like; inflorescences branching; sepals and petals spatulate, more than 5 times as long as broad **14. Corymborkis**
- Stems and rhizomes fleshy; inflorescences terminal, unbranched; sepals and petals less than 3 times as long as broad 16

16. Plants usually more than 50 cm tall; leaves distichous, plicate; lip bipartite, hypochile deeply saccate, lacking paired glands within, epichile fleshy and articulated **15. Epipactis**
- Plants usually less than 40 cm tall; leaves spirally arranged, fleshy; lip not bipartite, paired glands at the base within hypochile 17

17. Sepals free; flowers glandular-pubescent **12. Platylepis**
- Sepals united for half length; flowers glabrous **13. Cheirostylis**

18. Inflorescence produced and setting fruit before leaf develops **16. Nervilia**
- Inflorescence produced with leaves 19

19. Inflorescence terminal, produced between the terminal leaves on the stem or pseudobulb 20
- Inflorescence lateral from base of stem or pseudobulb 21

20. Column very short, enveloped by basal auricles of lip; lip flat, bearing 1 or 2 cushion-like calli **17. Malaxis**
- Column elongate, not enveloped by basal auricles of lip; callus fleshy, basal, entire or obscurely 2-lobed **18. Liparis**

21. Plants with pseudobulbs 22
- Plants lacking above-ground pseudobulbs 23

22. Pseudobulbs of 1 node, 1- to 3-leaved at apex **23. Oeceoclades**
- Plants with pseudobulbs of several nodes, 3- to many-leaved **25. Eulophia** (in part)

23. Lip lacking a spur; column very short, lacking a foot **24. Pteroglossaspis**
- Lip spurred or saccate at base; column elongate, with a foot **25. Eulophia** (in part)

24. Plants sympodial, often with pseudobulbs 25
- Plants monopodial, never pseudobulbous 28

25. Pseudobulbs of one node, 1- or 2-leaved at apex 26
- Pseudobulbs or stems of 2 or more nodes, usually many-leaved 27

26. Inflorescences 1-flowered; flowers with 8 pollinia **21. Stolzia**
- Inflorescences at least 2-flowered; flowers with 4 pollinia **22. Bulbophyllum**

27. Lip spurred; lateral sepals not forming a mentum with the column-foot; 2 types of root present: erect and sharp, spreading and branched **26. Graphorkis**
- Lip unspurred; lateral sepals forming a mentum with the column-foot; roots of 1 type only **20. Polystachya**

28. Flowers many, in a capitate head — **35. Ancistrorhynchus**
- Flowers solitary or in an elongate raceme — 29

29. Leaves bilaterally flattened and arranged in a fan — 30
- Leaves flat or terete — 31

30. Flowers campanulate, white, distichous; spur present — **31. Bolusiella**
- Flowers flat, pale yellow, whorled; spur absent — **19. Oberonia**

31. Rostellum strongly 3-lobed after removal of pollinia — **36. Angraecopsis**
- Rostellum 2-lobed after removal of pollinia — 32

32. Rostellum retuse; stipites very short and obscure, pollinia appearing to be directly attached to viscidia — **28. Angraecum**
- Rostellum elongate; stipites or stipe elongate — 33

33. Pollinia attached by a single stipe to a viscidium — 34
- Pollinia attached by separate stipites to 1 or 2 viscidia — 38

34. Lip strongly 3-lobed, sometimes obscurely — **37. Tridactyle**
- Lip entire or only obscurely lobed — 35

35. Inflorescence axis strongly zig-zag; spur with a distinct geniculate bend — **27. Calyptrochilum**
- Inflorescence not as above; spur straight or gently incurved — 36

36. Lip lanceolate-tapering; flowers white, stellate; spur elongate, more than 5 cm long — **32. Aerangis**
- Lip ovate to oblong; spur short, less than 2 cm long — **30. Diaphananthe** (in part)

37. Pollinia attached to separate viscidia; flowers pale yellow, green or rarely white, usually not stellate — **30. Diaphananthe** (in part)
- Pollinia attached to a single viscidium; flowers white, more or less stellate — 38

38. Lip entire; pollinia attached by oblanceolate stipites to a saddle-shaped viscidium — **34. Cyrtorchis**
- Lip 3-lobed, sometimes obscurely so; pollinia attached by 2 stipites to a single, oblong viscidium — **33. Rangaeris**

1. VANILLA *Mill.*

Scrambling vines or lianas. Stems elongate, green, bearing leaves or scales and adventitious roots opposite the leaves. Leaves, if present, fleshy or leathery, or reduced to scales. Inflorescences lateral, dense racemes with spirally arranged flowers and bracts. Flowers showy. Pedicel and ovary cylindrical. Sepals and petals more or less spreading, lanceolate to oblanceolate. Lip large, entire or obscurely 3-lobed, often bearing a tufted callus in the centre. Column elongate; pollinia mealy. Fruits bean-like. Seeds with a crustose testa.
A large genus of about 100 species widely distributed in the tropics of the Old and New Worlds. A single species is found in Ethiopia. *Vanilla* is the only representative of subfamily Vanilloideae found in Africa. Commercial vanilla (the Mexican *Vanilla planifolia*) is widely cultivated in the tropics for its pods which are cured and used as a flavouring. It is readily distinguished from the native species by its leafy stems and yellow-green flowers. There is no record of its cultivation in Ethiopia.

V. roscheri *Rchb.f.*

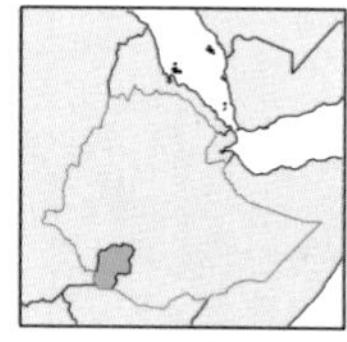
Vanilla roscheri

The specific epithet commemorates Albrecht Roscher, the German collector who discovered the type specimen in Zanzibar. It was described by H.G. Reichenbach in 1877.

Stems green, very long, occasionally branching, and with adventitious roots. Leaves reduced to green scales and drying brown. Inflorescences short but bearing up to 35 flowers in succession. Flowers white with an apricot or pink streak in the lip. Sepals and petals oblong-lanceolate to elliptic-lanceolate, up to 8 × 3.8 cm. Lip entire, obtuse, up to 8 × 4.5 cm, the callus of two rows of laciniate crests. Column up to 2.5 cm long. Fruits up to 17.5 cm long.

Habitat and distribution Open evergreen *Acacia* bushland and scrub at 1050 m in Gamo Gofa. Also in coastal Kenya, Tanzania, Zanzibar and Mozambique.

Conservation status Endangered in Ethiopia. Scattered but not threatened elsewhere.

Flowering period December.

Notes Closely related to and possibly conspecific with *Vanilla phalaenopsis* from the Seychelles and *V. madagascariensis* from Madgascar.

V. roscheri

2. HOLOTHRIX *Rich. ex Lindl.*

Plants small, terrestrial, tuber-bearing, with 1 or 2 sessile ovate or orbicular leaves. Inflorescence unbranched. Flowers often one-sided, sessile or shortly stalked. Sepals subequal, free from each other, often hairy. Petals usually longer (often much longer) than sepals, entire or divided in the upper part into 3 or more finger-like lobes. Lip similar to petals but broader with more lobes, spurred at the base, adnate to the column. Column very short; anther-loculi parallel, caudicles very short, viscidia small, naked; stigma sessile.

A genus of 55 species in tropical Africa, South Africa and tropical Arabia. Represented by seven species in Ethiopia.

Key

1	Plants flowering after the leaves have withered; peduncle and rhachis glabrous bearing lanceolate bracts along length	2
–	Plants leaf-bearing when in flower; peduncle pubescent, lacking bracts or rarely with a few along length	4
2	Lip entire; spur less than 2 mm long; petals blunt or rounded at apex; inflorescence a dense cylindric head	**5. H. squammata**
–	Lip with 1–3 slender apical lobes; spur 3–15 mm long; petals lobed at apex, midlobe slender, linear; inflorescence laxly many flowered, more or less secund	3
3	Spur 12–15 mm long; petals and lip margins papillose; petals with 3 linear lobes at apex	**3. H. praecox**
–	Spur 3.5–8.5 mm long; petals and lip margins glabrous; petals obscurely 3-lobed at apex, side lobes shorter than midlobe	**1. H. aphylla**
4	Basal leaf solitary; lip obscurely 3-lobed in basal half; midlobe obovate, pubescent	**7. H. unifolia**
–	Basal leaves 2; lip 3–7-lobed in apical half or if entire then lilac, blue or violet	5
5	Lip lilac, blue or violet, at least 4 times longer than sepals and petals	**4. H. brongniartiana**
–	Lip green, pinkish, dark red or purple, less than twice as long as sepals and petals	6
6	Lip 7-lobed, pubescent, pinkish, central 3 lobes blunt	**6. H. tridentata**
–	Lip 3-lobed, glabrous, green flushed with purple or violet, lobes acute	**2. H. arachnoidea**

H. brongniartiana

1. H. aphylla *(Forssk.) Rchb.f.*

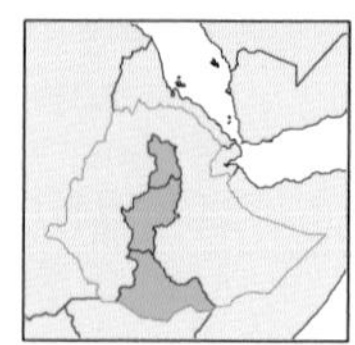

Holothrix aphylla

The specific epithet '*aphylla*' refers to the absence of leaves during the flowering period. Pehr Forsskål described it as *Orchis aphylla* in 1775 from a plant he collected in Yemen. H.G. Reichenbach transferred it to *Holothrix* in 1881.

Perennial herb, 6–27 cm tall. Leaves 2, basal, suborbicular or reniform, up to 3 × 4 cm, usually shrivelled at or about time of flowering. Inflorescence many-flowered, one sided, rather lax. Bracts lanceolate, shorter than the ovary. Flowers white, slightly tinged purple or bluish. Sepals lanceolate or ovate-lanceolate, 2.5–4 mm long. Petals narrowly ovate or oblong, 4–8 mm long, more or less 3-lobed at the apex, the midlobe often longer than the side lobes, linear, more or less papillose. Lip 2.5–6.5 mm wide, 3-lobed at the apex; spur tapering at the apex, 3.5–8.5 mm long.

Habitat and distribution In grazed grassland with scattered *Acacia* between 2000 and 2650 m in Welo, Shewa and Sidamo. Also in Yemen, Nigeria, Uganda, Kenya, DR Congo and Cameroon.

Flowering period December to February; March to April.

Conservation status Rare and vulnerable but possibly overlooked.

Notes *H. aphylla* differs from *H. praecox* by the spur being 3.5–8.5 mm rather than 12–15 mm long and the margins of petals glabrous rather than papillose.

2. H. arachnoidea *(A.Rich.) Rchb.f.*

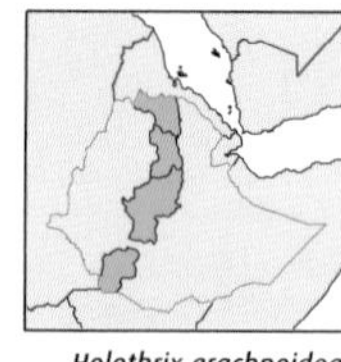

Holothrix arachnoidea

The specific epithet '*arachnoidea*' refers to the spider-like flowers. Achille Richard described it as *Peristylus arachnoideus* in 1840 from a plant Richard Quartin-Dillon collected on Mt. Seloda, near Adwa in Tigray. The name was changed to *H. arachnoidea* by H.G. Reichenbach in 1881.

Plant 15–70 cm tall. Leaves 2, lanceolate, very broadly ovate or orbicular, 2–6.5 × 1–4.5 cm, with long rather soft hairs. Inflorescence erect, densely many-flowered; scape densely long hairy. Bracts lanceolate, equalling the ovary, hairy. Flowers very small, secund, glabrous, green flushed with violet or purple; ovary glabrous or sparsely hairy. Sepals ovate, the laterals very oblique, 1.5–2.25 mm long, glabrous. Petals obliquely lanceolate to ligulate, 2–3 mm. Lip narrowly ovate,

H. arachnoidea

2.2–3.2 mm long, 3-lobed in the upper half or third; lobes nearly equal; spur conical-cylindrical, 1.5 mm long. Ovary glabrous or sparsely hairy.

Habitat and distribution In upland grassland, granite slopes covered by scrub and in *Juniperus procera* forest between 2500 and 3000 m in Tigray, Welo, Shewa and Gamo Gofa. Also in Kenya, Somalia, Sudan, Yemen and Saudi Arabia.

Flowering period August to October; December.

Conservation status Local but possibly overlooked.

Notes Differs from *H. tridentata* by the lip being 7-lobed rather than 3-lobed.

3. H. praecox *Rchb.f.*

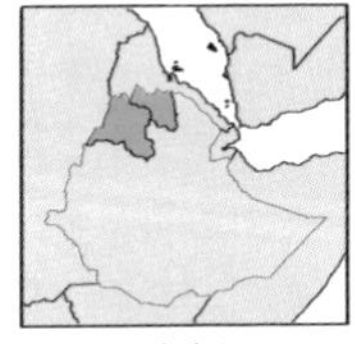

Holothrix praecox

The specific epithet '*praecox*' refers to the condition whereby the flowers develop before the leaves. H.G. Reichenbach described it in 1881 from a plant Georg Wilhelm Schimper collected in Semien, Gonder in northern Ethiopia.

Plant up to 30 cm tall. Leaves shrivelled at time of flowering. Inflorescence with laxly 10–16-flowered racemes. Bracts lanceolate, equalling or slightly longer than the ovary. Ovary 8 mm long. Dorsal sepal ovate-lanceolate, 4.6 × 1.5 mm. Lateral sepals triangular, 4 × 1.7 mm. Petals oblong with papillose margins 7.8 ×

2 mm, extended at apex into 3 linear lobes up to 2.2 mm long. Lip oblong, boat-shaped, margins papillose, 4.5 × 2.1 mm, 3-lobed; side lobes deflexed, sometimes bidentate; midlobe sometimes long toothed or filiform, up to 1.4 mm; spur cylindrical, 12–15 mm long.

Habitat and distribution In montane grassland between 2600 and 2800 m in Tigray and Gonder. Also in DR Congo.

Flowering period April.

Conservation status Rare and endangered.

Notes Differs from *H. aphylla* by the longer spur and the margins of petals papillose rather than glabrous.

4. H. brongniartiana *Rchb.f.*

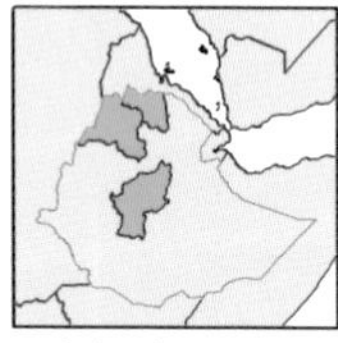

Holothrix brongniartiana

The specific epithet '*brongniartiana*' was given in honour of the French botanist Adolph Brongniart. H.G. Reichenbach described it in 1881 from plants Georg Wilhelm Schimper collected in Semien, Gonder.

Perennial herb 5–20 cm tall. Leaves 2, reniform-orbicular, 2–5 × 3–5 cm, glabrous. Scape hairy. Bracts lanceolate, shorter than the pedicellate ovary, nearly glabrous. Flowers lilac or pale mauve. Sepals ovate or broadly lanceolate, the laterals very oblique, 2.5–4.5 mm long, almost glabrous. Petals obliquely lanceolate, 3.5–4.5 mm long. Lip much longer than other petals and sepals, oblanceolate or narrowly elliptical, rounded or very shortly 2-lobulate at the very tip, 7.5–10 mm long; spur narrowly conical, 3.5–4 mm long. Ovary hairy.

Habitat and distribution On steep slopes with *Erica* and *Juniperus* shrubs or under *Carissa edulis*, *Rosa abyssinica*, *Maesa lanceolata*, upland grassland, often among rocks between 2400 and 3500 m in Tigray, Gonder and Shewa. Also in Uganda, Kenya, Tanzania, Zambia and Malawi.

Flowering period April to September.

Conservation status Vulnerable.

Notes Differs from *H. arachnoidea* and *H. tridentata* by the lip being at least 4 times longer than the sepals and petals rather than less than twice as long.

5. H. squammata *(A.Rich.) Rchb.f.*

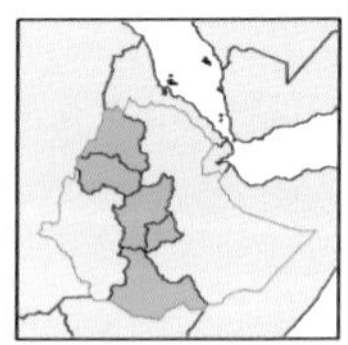
Holothrix squammata

The specific epithet '*squammata*' refers to the scale-like sterile bracts on the stem. Described as *Peristylis squammatus* by Achille Richard in 1851 from a plant Georg Wilhelm Schimper collected in Semien, Gonder. The name was changed to *H. squammata* by H.G. Reichenbach in 1881.

Perennial herb 5–22 cm tall. Leaves 2, generally shrivelled at the time of flowering, ovate or orbicular-reniform, up to 4 × 4 cm, glabrous. Inflorescence dense, many-flowered. Bracts lanceolate, as long as or shorter than the flowers. Flowers white. Sepals triangular-ovate, 2–4 mm long, glabrous. Petals elliptical, rounded at the apex, 3–5 mm long. Lateral sepals elliptical to ovate-elliptical, rounded at the apex, 3–4.5 × 2–3.5 mm. Lip entire, ovate-obovate, 4.5 × 3.5 mm; spur narrowly conical, 1.3–2.5 mm long. Ovary glabrous, 1 cm long.

Habitat and distribution In upland grassland or moorland, dry open places and in shade of *Juniperus procera* between 2400 and 3300 m in Gonder, Gojam, Shewa, Arsi and Sidamo. Also in Sudan and Uganda.

Flowering period December to May.

Conservation status Locally common and possibly overlooked.

Notes Differs from *H. aphylla* and *H. praecox* by its dense-flowered inflorescence and its shorter spur.

6. H. tridentata *(Hook.f.) Rchb.f.*

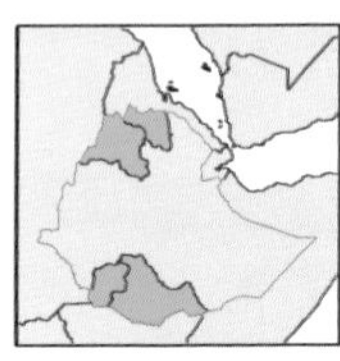
Holothrix tridentata

The specific epithet '*tridentata*' refers to the 3-lobed lip. Described as *Peristylus tridentatus* by Joseph Hooker in 1864 from a plant Gustav Mann collected on Mt. Cameroon. The name was changed to *H. tridentata* by H.G. Reichenbach in 1881.

Perennial herb 6–14 cm tall. Leaves 2, very broadly ovate or orbicular, up to 3 × 3 cm, with ciliolate margins. Scape pilose, more pubescent above. Inflorescence erect, 4–12-flowered. Bracts triangular-ovate, hairy, about equalling the ovary. Flowers very small, secund, glabrous, green flushed with violet or purple, lip pinkish; ovary hairy. Sepals ovate, the laterals somewhat oblique, 2–2.8 mm long, glabrous. Petals broadly obovate, acuminate or obscurely tridentate;

H. tridentata

side lobes reduced and obtuse, 2.4–3.1 mm long. Lip ovate, 3.2–4.5 mm long, 7-lobed in the upper half or third; central lobes blunt, outer lobes acute; spur conical, 1.5 mm long.

Habitat and distribution In grassland and on steep slopes between 2100 and 2250 m in Tigray, Gonder, Sidamo and Gamo Gofa. Also in Sudan and Cameroon.

Flowering period August and September.

Conservation status Endangered.

Notes Differs from *H. arachnoidea* by the lip being 3-lobed and pubescent rather than 7-lobed.

7. H. unifolia *(Rchb.f.) Rchb.f.*

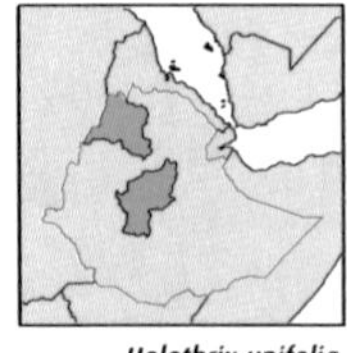

Holothrix unifolia

The specific epithet '*unifolia*' refers to the solitary leaf. H.G. Reichenbach described it as *Deroemeria unifolia* in 1855 from a plant collected by Georg Wilhelm Schimper in Semien, Gonder. He transferred it to *Holothrix* in 1881.

Perennial herb 10–20 cm tall. Leaf solitary, very broadly ovate or orbicular, 2.5–5.5 × 3–6.5 cm. Inflorescence erect, densely many-flowered with hispid scape. Bracts triangular-ovate, about equalling the ovary. Flowers white with red or purple markings. Dorsal sepal ovate, 2–3.2 × 1–1.3 mm. Lateral sepals somewhat oblique, ovate, 2.3–3 × 1.3–1.5 mm. Petals ovate, more or less pubescent, exceeding the sepals, 3.8–4.2 × 1–1.6 mm. Lip oblong in outline, pubescent, 3.8–4.5 × 2.5–2.7 mm, obscurely and minutely 3-lobed, rarely 5–7-lobed; spur cylindrical, 2.5–3 mm long.

Habitat and distribution In open, dry grasslands sometimes with scattered shrubs between 2500 and 2900 m in Gonder and Shewa. Unknown elsewhere.

Flowering period October to January.

Conservation status Rare and critically endangered.

Notes Differs from *H. arachnoidea*, *H. tridentata* and *H. brongniartiana* by having a solitary basal leaf rather than two.

H. unifolia

3. BRACHYCORYTHIS *Lindl.*

Plants terrestrial, tuber-bearing, up to 1 m high. Stems with often numerous and overlapping leaves, which are usually lanceolate. Inflorescence terminal. Flowers few to numerous, white, yellow, pink or various shades of mauve or purple, often spotted darker. Bracts leafy, the lower exceeding the flowers. Sepals free, the lateral ones spreading and oblique. Petals usually adnate at the base to the side of the column. Lip projecting forwards, rarely deflexed; basal part (hypochile) boat-shaped, sac-like or spurred; upper part (epichile) flattened, entire or 2- or 3-lobed. Column erect, rather slender; anther-loculi parallel; caudicles usually very short, viscidia naked; stigma hollowed out, rostellum middle lobe small, erect, folded, the side lobes often fleshy, surrounding the viscidia.

A genus of 32 species in tropical and South Africa, and tropical Asia. Only three species are known from Ethiopia.

Key

1	Leaves completely covered by a velvety pubescence	**2. B. pubescens**
-	Leaves glabrous	2
2	Lip with a small conical callus at base of epichile; dorsal sepal 4–5 mm long	**1. B. buchananii**
-	Lip lacking a callus at base of epichile; dorsal sepal 5–14 mm long	**3. B. ovata** subsp. **schweinfurthii**

1. B. buchananii (*Schltr.*) *Rolfe*

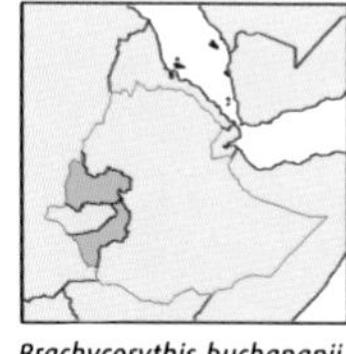

Brachycorythis buchananii

The specific epithet '*buchananii*' was given in honour of John Buchanan, who collected the type specimen in Malawi. Rudolf Schlechter described it as *Platanthera buchananii* in 1897.

Plant slender, 20–55 cm tall. Leaves numerous, lanceolate, up to 4.5 × 1 cm, decreasing in size up the stem. Inflorescence slender, densely many-flowered, up to 14 × 2 cm. Bracts lanceolate, the lower equalling the flowers. Flowers pink, mauve or purple; ovary with pedicel 5–7 mm long. Sepals lanceolate-elliptical, the laterals slightly oblique, 4–6 mm long. Petals obovate or semi-orbicular, a little longer than the sepals and much broader, many-veined. Lip bipartite; hypochile boat-shaped with triangular sides, 1.5–2.5 mm long; epichile

B. buchananii

kidney-shaped or elliptical, 3-lobed with the side lobes longer than middle lobe, bearing a small callus just in front of the hypochile, 2.5–3.5 × 4–4.5 mm. Column 2 mm long.

Habitat and distribution In savannah country with scattered trees and forest along rivers and creeks between 1250 and 1500 m in Kefa and Wellega. Also in Nigeria, DR Congo, Angola, Uganda, Kenya, Tanzania, Malawi, Zambia and Zimbabwe.

Flowering period July to September.

Conservation status Rare in Ethiopia but locally common elsewhere.

Notes Differs from *B. ovata* by its smaller flowers.

2. B. pubescens *Harv.*

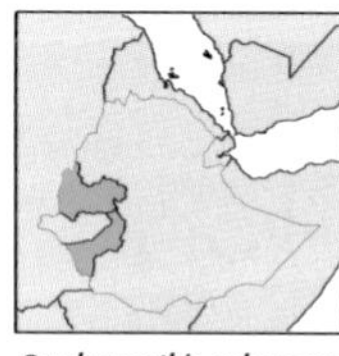

Brachycorythis pubescens

The specific epithet '*pubescens*' refers to the velvety hairy nature of the leaves, inflorescence and flowers. William Harvey described it in 1859 from a plant John Sanderson collected near Durban in Natal, South Africa.

Plant 25–80 cm tall. Leaves numerous, lanceolate or broadly lanceolate, those in middle of stem up to 6 × 2.5 cm, decreasing in size up the stem, densely velvety hairy. Inflorescence densely many-flowered, up to 35 × 2–6 cm. Bracts similar to leaves, the lower ones usually longer than the flowers, velvety hairy. Flowers various shades of pink or purple, often with an orange centre; pedicel and ovary 1–2 cm long, velvety. Dorsal sepal elliptical, 4.5–7.5 mm long; laterals obliquely ovate or elliptical; all sepals pubescent outside. Petals very obliquely elliptical or oblong, slightly shorter than the dorsal sepal, with a few hairs on the outside. Lip hypochile shortly bowl-shaped, 2–3 mm long from back to front, the margins angular; epichile bent downwards in a knee-like manner from its attachment, broadly wedge-shaped or fan-shaped, 3-lobed at the apex, 5–10 × 6–14 mm, the lobes triangular, often rounded, the middle one smaller than or equalling the side lobes.

Habitat and distribution In grassland or open woodland between 1500 and 1800 m in Kefa and Wellega. Widespread in tropical Africa and South Africa.

Flowering period June to September.

Conservation status Rare and endangered in Ethiopia but locally common elsewhere.

Notes Differs from *B. buchananii* and *B. ovata* subsp. *schweinfurthii* by the leaves being covered by velvety pubescence all over rather than being glabrous.

B. pubescens

3. B. ovata *Lindl.* subsp. schweinfurthii *(Rchb.f.) Summerh.*

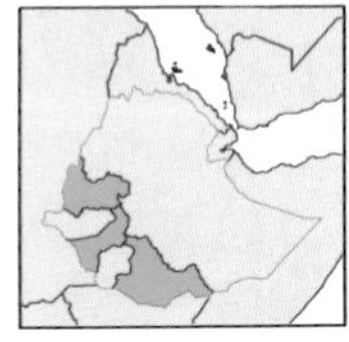

Brachycorythis ovata

The subspecific epithet '*schweinfurthii*' is given in honour of Georg August Schweinfurth, who collected the type specimen in Sudan. H.G. Reichenbach described it as *B. schweinfurthii* in 1878. Its status was changed to a subspecies of *B. ovata* by Victor Summerhayes in 1955.

Plant 20–100 cm tall. Leaves very numerous, narrowly to broadly lanceolate, the middle ones up to 8 × 2.5 cm broad, decreasing in size up the stem. Inflorescence densely to laxly many-flowered, up to 35 cm × 3–6 cm in diameter. Bracts leaf-like, lanceolate, the lower ones longer than the flowers. Flowers lilac with darker spotting, various shades of purple and mauve, often with an admixture of white; pedicel and ovary 1–2 cm long. Dorsal sepal elliptical, 5–9 mm long; laterals obliquely ovate, 6.5–11 mm long. Petals obliquely oblong-ovate with a marked rounded auricle in front, truncate or toothed at the apex, about as long as the dorsal sepal. Lip bipartite; hypochile boat-shaped, somewhat curved, 4–6 mm long; epichile broadly obovate, 3-lobed in the upper part, with a keel running along the centre into the triangular middle lobe, the side lobes more or less incurved, shorter than, equalling or longer than the middle lobe.

Habitat and distribution In open secondary scrub, pasture, *Combretum* woodland and *Eucalyptus*-planted grassland between 1600 and 2000 m in Wellega, Sidamo and Kefa. Also in Senegal, Ivory Coast, Nigeria, Cameroon, DR Congo, Sudan, Uganda, Kenya, Tanzania.

Flowering period May and June.

Conservation status Vulnerable in Ethiopia but locally common elsewhere.

Notes Differs from *B. buchananii* by its much larger flowers.

4. CYNORKIS *Thouars*

Plants terrestrial, tuber-bearing. Stems often glandular-pubescent. Leaves few or solitary, almost all radical, the ones on the stem small or sheath-like. Flowers few or numerous in a lax or dense terminal raceme, usually resupinate but rarely not so, usually pink or mauve, less frequently orange, rarely white. Sepals free or slightly adnate to the lip, the laterals spreading, the dorsal often forming a helm with the 2 petals. Lip free, entire or 3–5-lobed, usually larger than the tepals, spurred at the base. Column short and broad; androclinium erect or sloping; anther-loculi parallel, canals short or long and slender, caudicles slender, viscidia 2, rarely 1, naked, auricles distinct; stigmatic processes oblong, papillose, usually united to the rostellum lobes; rostellum prominent, several-lobed, the side lobes usually elongated, the middle lobe often large, projecting forward. Capsules oblong or fusiform.

A genus of about 125 species, mostly natives of Madagascar and the Mascarene Islands; 17 species in Africa. Only two species are known from Ethiopia.

Key

1 Lip ligulate; inflorescence pyramidal, densely many-flowered; sepals and petals 3–5 mm long; spur 3–5 mm long, tapering-incurved **1. C. anacamptoides**

– Lip ovate or broadly lanceolate, 3-lobed in middle; inflorescence cylindrical; sepals and petals 5–8 mm long; spur 6–9 mm, cylindrical, slightly clavate, straight **2. C. kassneriana**

C. kassneriana

1. C. anacamptoides *Kraenzl.*

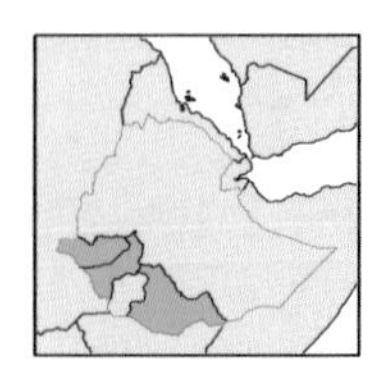

Cynorkis anacamptoides

The specific epithet '*anacamptoides*' refers to the resemblance of the inflorescence to that of *Anacamptis pyramidalis*, the European pyramidal orchid. Fritz Kraenzlin described it in 1895 from a plant Stuhlmann collected in the Ruwenzori Mountains in DR Congo.

Plant 10–60 cm tall. Stem slender, usually glandular-pubescent in the upper part, bearing 2–6 leaves near the base and several smaller sheath-like ones above. Leaves lanceolate or lanceolate-elliptic, the lowest ones often stalked, acute, the largest 2–13 cm × 8–18 mm, glabrous. Inflorescence densely many-flowered, 2.5–20 cm long. Bracts lanceolate, glabrous or more or less glandular-pubescent, the lower ones longer then the ovary. Flowers pink, mauve or purple. Dorsal sepal ovate, convex, 3 mm long, forming a hood with the petals; laterals spreading, obliquely and broadly oblanceolate, 3.5–5 mm long. Petals lanceolate or oblanceolate, curved, 3–4 mm long. Lip ligulate or narrowly wedge-shaped, 3–5 × 1 mm; spur incurved-cylindrical, slightly swollen in apical half, 3–5 mm long.

Habitat and distribution In upland moorland or upland grassland, bogs with *Sphagnum* or tall grass, swamps, marshes by streams or in open places in upland rainforests between 1900 and 3000 m in Sidamo, Illubabor and Kefa. Also in Cameroon, Bioko, DR Congo, Uganda, Tanzania, Angola, Zambia, Malawi and Zimbabwe.

Flowering period June to November.

Conservation status Locally common throughout its range but rare in Ethiopia.

Notes Differs from *C. kassneriana* by its pyramidal inflorescence of smaller flowers.

C. anacamptoides

2. C. kassneriana *Kraenzl.* subsp. kassneriana

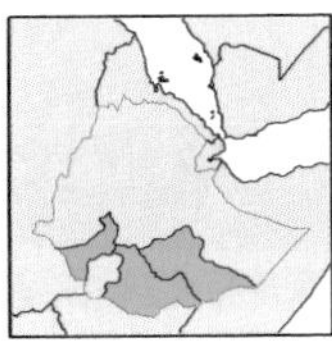

Cynorkis kassneriana

The specific epithet '*kassneriana*' is given in honour of the German collector Theodore Kassner from whose collection from the Ruwenzori in DR Congo Fritz Kraenzlin described it in 1914.

Plant 15–50 cm tall. Stem slender, glandular-hairy. Leaf erect or spreading, lanceolate, elliptical-lanceolate or oblanceolate, 4–20 × 1–4 cm, glabrous. Inflorescence laxly or rather densely 4–22-flowered, 3–10 cm long; rhachis glandular-hairy. Bracts lanceolate, shorter than the ovary, glandular-ciliate. Flowers pinkish-purple or mauve; pedicel and ovary 1–2 cm long. Ovary glandular-hairy. Dorsal sepal very convex, ovate, hooded at the apex, 5–7 mm long; laterals spreading, obliquely ovate or lanceolate-ovate, 6.5–8 mm long; all more or less sparsely glandular-hairy outside. Petals obliquely lanceolate, 4.5–7.5 mm long, adnate at the inner base to the column. Lip ovate or broadly lanceolate in outline, more or less distinctly 3-lobed at the middle, 5–8.5 × 2.5–5 mm, the middle lobe triangular and much larger and longer than the side lobes; spur descending, cylindrical, 6–9 mm long.

C. kassneriana

Habitat and distribution On mossy banks by streams, along rocks, on forest floor, or epiphytic on moss-covered branches in upland rainforest between 2000 and 2580 m in Bale, Sidamo and Kefa. Also in DR Congo, Uganda, Tanzania, Malawi, Zambia, Zimbabwe and South Africa (Transvaal).

Flowering period August to October.

Conservation status Rare and vulnerable in Ethiopia but locally common elsewhere.

Notes Differs from *C. anacamptoides* by its larger flowers.

5. HABENARIA *Willd.*

Terrestrial, or rarely epiphytic, herbs with elongated fleshy or tuberous roots. Stems unbranched, sometimes very short. Leaves variously arranged along the stem, sometimes with 1 or 2 radical and appressed closely to the ground, the cauline leaves sometimes sheath-like. Inflorescence terminal, 1–many-flowered. Flowers usually resupinate, but in a few species not so, usually white and/or green, rarely yellow, orange or pink. Sepals usually free, the laterals spreading, the dorsal often forming a helm with the 2 petals. Petals often adherent to the dorsal sepal, entire or variously divided, often 2-lobed nearly to the base. Lip usually slightly adnate at the base to the column, the free part entire or variously divided or lobed, spurred at the base; spur short, sac-like to long and slender. Column tall or short, slender or thickened; anther upright or reclinate, the loculi adjacent and parallel with a narrow connective, or separated from one another by a much broadened filament and more or less divergent, canals short or much elongated, adnate to the lateral lobes of the rostellum, auricles (staminodes) sometimes elongated or 2-lobed, usually rugose; pollinaria 2, each with sectile pollinium, short or elongated caudicle and rather small naked viscidium; stigmatic processes distinct, shortly club-shaped to very long with capitate or club-shaped apices, usually free, but sometimes united in the lower part to the rostellum, the rostellum side lobes divergent, short or long, middle lobe tall and overtopping the anther to short and very blunt or scarcely developed. Capsules oblong or fusiform.

A genus of approximately 600 species, distributed throughout tropical and subtropical regions. Twelve sections including 47 species are found in Ethiopia.

Key to sections for Ethiopian species

1	Leaves 1 or 2, basal, appressed to the ground	**Sect. Diphyllae**
–	Leaves borne along stem, not appressed to the ground	2
2	Petals densely hairy or longly ciliate	**Sect. Trachypetalae**
	Petals glabrous or only sparsely hairy	3
3	Petals entire	4
–	Petals bipartite	7

H. distantiflora

4	Spur short, 0.5–2 mm, globose; stigmatic processes less than 1 mm long	**Sect. Pseudoperistylus**
–	Spur usually much longer than 2 mm, cylindrical or inflated at the apex but not globose; stigmatic processes usually more than 1.5 mm long	5
5	Inflorescence cylindrical, densely many-flowered	**Sect. Commelynifoliae**
–	Inflorescence slender, narrow and/or laxly flowered	6
6	Side lobes of lip divided into narrow segments or if not divided then over 5 mm broad	**Sect. Multipartitae**
–	Side lobes of lip entire and less than 2 mm broad	**Sect. Chlorinae**
7	Dorsal sepal strongly reflexed	**Sect. Replicatae**
–	Dorsal sepal more or less erect	8
8	Leaves oblanceolate; anther canals less than 1 mm long; stigmatic processes 1.5–2.5 mm long	**Sect. Pentaceras**
–	Leaves linear to lanceolate to suborbicular but not oblanceolate; anther canals longer than 1 mm; stigmatic processes longer than 3 mm	9
9	Petals papillate, pubescent or ciliate at least in part	10
–	Petals glabrous	11
10	Stigmas curved downwards over base of lip	**Sect. Mirandae**
–	Stigmas more or less porrect	**Sect. Cultratae**
11	Anther canals 4–20 mm long	**Sect. Ceratopetalae**
–	Anther canals less than 3 mm long	**Sect. Macrurae**

Section **Chlorinae** *Kraenzl.*

A section in which most species have small green flowers in which the petals are entire and the lip has entire side lobes less than 2 mm long.

Key

1	Length of midlobe of lip 7 or more times width, only slightly wider than side lobes	**3. H. filicornis**
–	Length of midlobe of lip less than 4 times width, much wider than side lobes	2
2	Leaves 2, well developed at base of stem; apical half of bracts drying with dark staining; lip midlobe length exceeding side lobes	**1. H.distantiflora**
–	Leaves well developed all along stem; bracts of uniform colour; lip midlobe shorter than side lobes	**2. H. bracteosa**

1. H. distantiflora *A.Rich.*

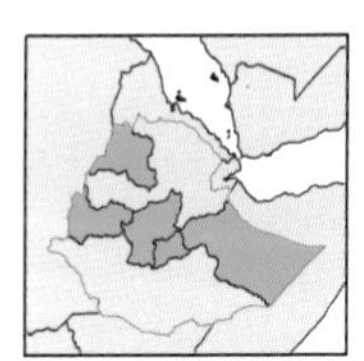
Habenaria distantiflora

The specific epithet '*distantiflora*' refers to the laxly flowered inflorescence. Achille Richard described this species in 1851 from a plant collected near Shire, Tigray in northern Ethiopia by Quartin-Dillon.

Plant 15–50 cm tall. Leaves 4–7, very unequal, 1–2 at base of stem, lanceolate-oblong, up to 13 × 1.3–2 cm, the upper ones much smaller, similar to the bracts. Inflorescence slender, up to 16 cm long, rather lax, up to 20-flowered. Bracts longer than the flowers, apical half with darker coloration. Flowers apparently in a single spiral row, green or yellowish-green with brown markings; pedicel with ovary 1 cm long. Dorsal sepal incurved, 2.7–3.3 × 2.1–2.4 mm. Lateral sepals deflexed, 3–4 mm long; all sepals ciliate. Petals broadly lanceolate, back margin infolded, equalling the dorsal sepal and forming a helm with it. Lip 3-lobed nearly to the base, the claw 0.5–1 mm long; lobes oblong, the midlobe equalling or longer and broader than the spreading side lobes; spur pendent, slender, slightly swollen in the distal half, 10–17 mm long.

Habitat and distribution In upland grassland among bracken, *Juniperus procera* or *Erica arborea* between 2000 and 3800 m in Tigray, Gonder, Shewa, Arsi, Harerge and Wellega. Also in Fernando Po, Cameroon, DR Congo, Sudan, Kenya, Uganda and Yemen.

Flowering period July to September.

Conservation status Locally common.

Notes Differs from *H. bracteosa* by the shorter lateral sepals and spur.

2. H. bracteosa *A.Rich.*

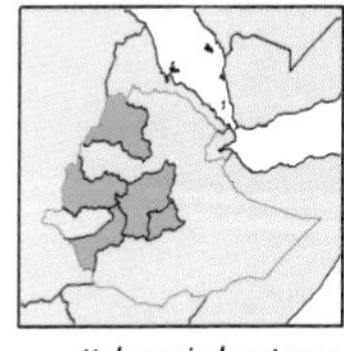
Habenaria bracteosa

The specific epithet '*bracteosa*' refers to the large bracts which exceed the flowers. Achille Richard described it in 1851 from a plant collected in Semien, northern Ethiopia by Georg Wilhelm Schimper.

Plant 15–95 cm tall. Leaves 5–10, the lowest 2 sheath-like, the middle 3–5 lanceolate or oblong-lanceolate, 10–30 × 0.5–3 cm, the uppermost smaller but similar, bract-like. Inflorescence rather densely many-flowered, 5–50 cm long. Bracts narrowly lanceolate, the lower ones longer than the flowers. Flowers green or yellowish-green with yellow anther;

pedicel and ovary 1 cm long. Dorsal sepal ovate or narrowly ovate, 2.5–4.5 mm long. Lateral sepals obliquely and broadly oblong-lanceolate, 3.5–5.5 mm long. Petals very obliquely triangular-ovate, equalling the dorsal sepal but a little broader. Lip deflexed, 3–8 mm long, 3-lobed from a short broad claw; lobes oblong or ligulate, the side lobes a little longer or equalling the midlobe; spur slender, more or less incurved, 15–28 mm long.

Habitat and distribution In grassy glades in mountain forest, especially by streams, and in damp places in heath and other areas above forest between 1450 and 3600 m in Gonder, Shewa, Arsi, Kefa and Wellega. Also in Fernando Po, Cameroon, Sudan, Uganda, Kenya and Tanzania.

Flowering period July to September.

Conservation status Rare and vulnerable.

Notes Differs from *H. distantiflora* by the lateral sepals being 3.5–5.5 mm long rather than 3–4 mm long and the spur being 15–28 mm long rather than 10–17 mm long.

3. H. filicornis *Lindl.*

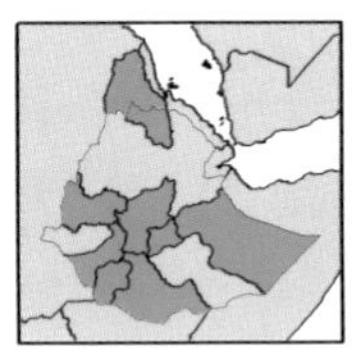

Habenaria filicornis

The specific epithet '*filicornis*' refers to the thread-like lobes of the lip. John Lindley described this species in 1835 from a plant collected in Ghana by Peter Thonning. The species was described as *H. tridactyla* by Achille Richard in 1851, from a plant collected near Adwa, Tigray, by Richard Quartin-Dillon.

Plant 15–50 cm tall. Stem slender, with scattered leaves. Leaves 3–9, the lowest 1 or 2 sheath-like, the remainder lanceolate or elliptical-lanceolate, the largest 3–11 × 0.7–1.8 cm, the others decreasing in size up the stem. Inflorescence 5–19 cm long, rather lax, up to 20-flowered. Bracts lanceolate, shorter than the flowers. Flowers curved outwards, green; ovary 8–11 mm long; pedicel 3–6 mm long. Dorsal sepal elliptical or ovate-elliptical, convex, 2.5–5.5 × 2.1–2.6 mm. Lateral sepals deflexed, obliquely semi-ovate, 3.5–7 × 1.9 –2.2 mm. Petals very obliquely lanceolate, as long as the dorsal sepal and adherent to it. Lip divided almost to the base, 5–13 mm long, the undivided part less than 1 mm long; lobes linear, rather fleshy, more or less incurved, the midlobe a little longer and broader than the side lobes; spur slender, slightly swollen towards the apex, 20–30 mm long.

H. filicornis

Habitat and distribution In open grassy slopes, coarse grassland with herbs and shrubby thickets or secondary bushland between 1000 and 2000 m in Tigray, Shewa, Arsi, Harerge, Sidamo, Wellega, Gamo Gofa and Kefa and in Eritrea. Also in Ghana, Ivory Coast, Nigeria, DR Congo, Tanzania, Uganda, Angola, Malawi, Zambia and Zimbabwe.

Flowering period April; July to September.

Conservation status Locally common.

Notes Differs from *H. bracteosa* and *H. distantiflora* by the midlobe being 7 or more times longer than wide.

Section **Pseudoperistylus** *P.F.Hunt*

A section in which most species have tiny flowers with entire petals and a trilobed lip with a basal callus and a short globose spur 0.5–2 mm long and stigmatic processes less than 1 mm long.

Key

1	Lip with two calli, one at base, the other on midlobe; leaves more or less basal; sepals and petals ciliate	**5. H. montolivaea**
–	Lip with a single central or apical callus; leaves spread along stem; sepals and petals glabrous	2
2	Lip side lobes longer than the midlobe, falcate-lanceolate; callus on lip midlobe; plants usually more than 18 cm tall	**6. H. petitiana**
–	Lip side lobes equalling midlobe, very short, triangular; callus on middle of lip; plants usually less than 15 cm tall	**4. H. lefebureana**

4. H. lefebureana *(A.Rich.) Th.Dur. & Schinz*

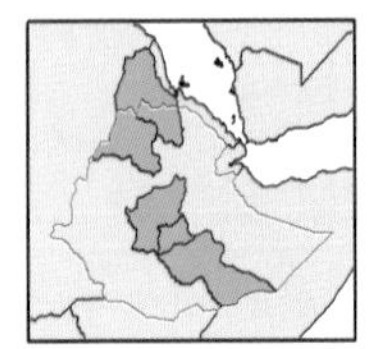

Habenaria lefebureana

The specific epithet '*lefebureana*' commemorates the French botanist Louis Victor Lefèvre. Achille Richard described it as *Peristylis lefebureanus* in 1840 from a plant Richard Quartin-Dillon and Antoine Petit collected in Adwa, Tigray.

Plant 8–40 cm tall. Stem leafy. Leaves up to 4.6 × 0.7–2.6 cm, elliptic to elliptic-ovate, largest in middle of stem. Inflorescence 2–5 × 0.8–1.3 cm, cylindrical or rarely pyramidal, densely many-flowered. Bracts 6–8 mm long, ovate to lanceolate. Flowers small, fragrant, white. Dorsal sepal 3–4 × 1.6–1.7 mm, elliptic to elliptic-ovate; lateral sepals 3.5–4 × 1.5 mm, similar to dorsal sepal. Petals 2.5–3 × 1.2–1.6 mm, oblique at base, elliptic. Lip 2.5–3 × 2.3–2.5 cm, subquadrate, 3-lobed at apex; lobes 0.5–0.7 mm long, subtriangular; callus fleshy, extending onto midlobe; spur 0.5 mm long, obscure.

Habitat and distribution In short grass on rocky slopes and in scrub between 2600 and 3800 m in Tigray, Gonder, Shewa, Arsi and Bale and in Eritrea. Also in Yemen.

Flowering period April in Bale; July to September elsewhere in Ethiopia.

Conservation status Locally common but vulnerable in Ethiopia.

Notes Differs from *H. petitiana* by being a much smaller plant with the side lobes of the lip equalling the midlobe.

H. lefebureana

5. H. montolivaea *Kraenzl. ex Engl.*

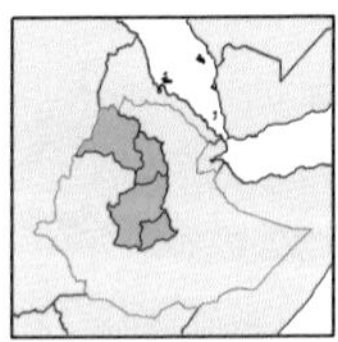
Habenaria montolivaea

The specific epithet '*montolivaea*' refers to the Mount of Olives, the name given to the type locality by Georg Wilhelm Schimper. Adolph Engler described it in 1892 using a name given by Fritz Kraenzlin based on a plant Schimper collected in Debre Tabor, Gonder.

Plant 10–20 cm tall. Stem erect, leafy along length. Leaves 4–7, the basal 3 spreading, obovate, the largest 2–4 × 1–3 cm, the uppermost much smaller, bract-like. Inflorescence up to 4 cm long, densely or subdensely many-flowered. Bracts lanceolate,the lower ones often longer than the flowers. Flowers white or greenish white; pedicel and ovary 4–6 mm long. Dorsal sepal ovate, 2.5–2.8 × 1.5–1.9 mm. Lateral sepals spreading, oblong or oblong-elliptical, 2.5–2.6 × 1.5–1.7 mm. Petals triangular or obliquely curved, ciliate, 2–2.6 × 0.8–1.7 mm. Lip 3-lobed in apical half, 2.5–3 × 3.5–4.2 mm, with a large fleshy callus or keel at base and apex; side lobes shorter or equalling midlobe; midlobe triangular, short; spur subglobose, 1.2–1.9 mm long.

Habitat and distribution In open grassland between 1000 and 2600 m in Gonder, Welo, Shewa and Arsi. Unknown elsewhere.

Flowering period April; July to September.

Conservation status Local and vulnerable.

Notes Differs from *H. lefebureana* and *H. petitiana* by the lip having two calli on its upper surface and the petals being ciliate.

6. H. petitiana *(A.Rich.) Th.Dur. & Schinz*

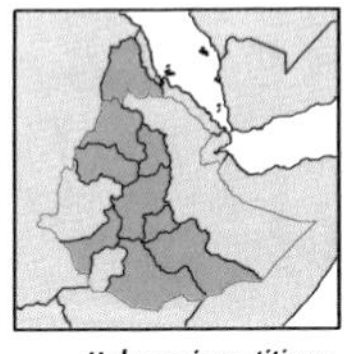
Habenaria petitiana

The specific epithet '*petitiana*' is given in honour of Antoine Petit, one of the collectors of the type specimen. It was collected in Tigray, northern Ethiopia. Achille Richard described it as *Peristylus petitiana* in 1840. Théophile Durand and Hans Schinz changed the name to *Habenaria petitiana* in 1895.

Plant up to 1 m tall. Stem erect, leafy along its whole length. Leaves 7–14, ovate or broadly lanceolate, the largest 3–9 × 1.5–4.3 cm, the uppermost often much smaller, bract-like. Inflorescence slender, 7–28 cm long but very rarely exceeding 20 cm, densely or rather densely many-flowered. Bracts lanceolate, the lower ones often longer than the flowers. Flowers fragrant,

Habenaria lips and spurs

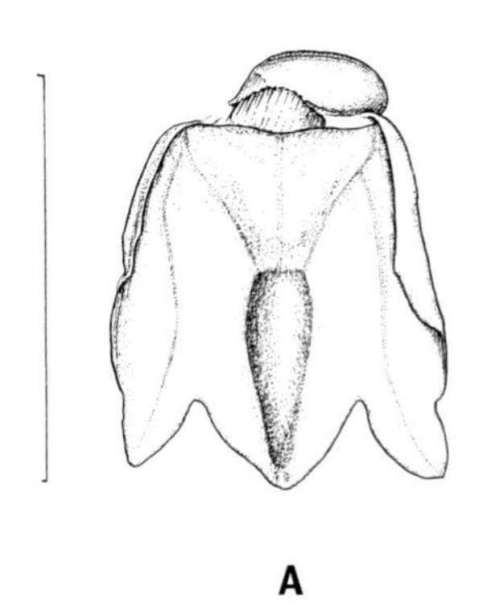

A

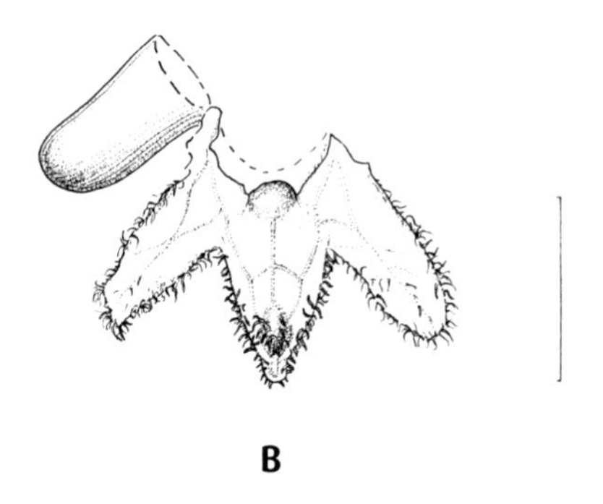

B

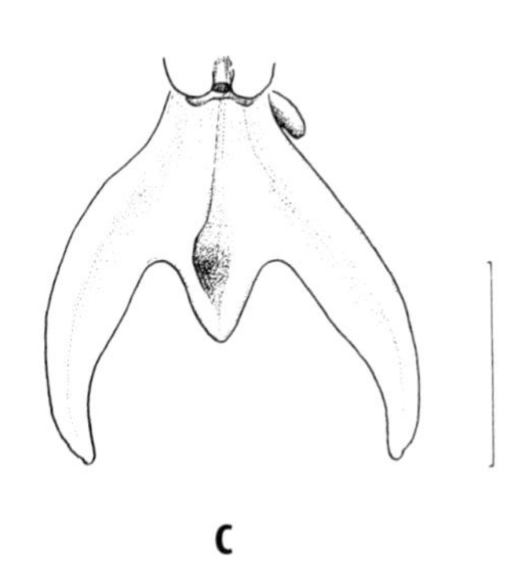

C

A *H. lefebureana;* **B** *H. montolivaea;* **C** *H. petitiana.* Scale bar = 2mm.

H. petitiana

green or yellow-green, glabrous; ovary almost sessile, 4–6 mm long. Dorsal sepal ovate or lanceolate-ovate, 2.5–4 mm long. Lateral sepals spreading, oblong or oblong-elliptical, a little longer but narrower than the dorsal sepal. Petals obliquely oblong, obtuse, 2–3.5 × 1–1.5 mm. Lip broadly cuneate or almost fan-shaped, 2.5–4 mm long and nearly as broad, 3-lobed in the apical half; lobes more or less divergent, triangular or ligulate-triangular, the laterals often rather longer than the midlobe, the latter with a distinct median keel; spur almost globose, just exceeding 1 mm in length.

Habitat and distribution In short grass, among scrub or at edges of forest, submontane or riverine between 1700 and 3150 m in Tigray, Gonder, Gojam, Welo, Shewa, Arsi, Bale, Sidamo and Kefa and in Eritrea. Also in DR Congo, Kenya, Tanzania, Uganda, Malawi and Zambia.

Flowering period July to September.

Conservation status Locally common in Ethiopia, widespread and locally common elsewhere.

Notes Differs from *H. lefebureana* by the side lobes of lip being longer than the midlobe rather than equal to it.

Section **Commelynifoliae** *Kraenzl.*

A section in which most species have cauline leaves like a *Commelina*, a densely flowered cylindrical inflorescence, entire petals and a 3-lobed lip.

Key

1	Spur deeply bilobed at the apex	**7. H. platyanthera**
–	Spur not deeply bilobed at the apex	2
2	Spur more than 25 mm long; side lobes of lip absent or less than a fifth length of midlobe	3
–	Spur 1.5–3 mm long; side lobes of lip half the length of the midlobe	**9. H. peristyloides**
3	Lip 3-lobed; spur 25–30 mm long	**8. H. epipactidea**
–	Lip entire; spur 30–65 mm long	**10. H. zambesina**

7. H. platyanthera *Rchb.f.*

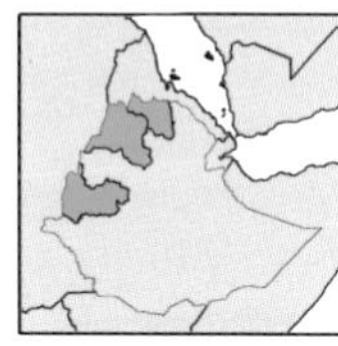
Habenaria platyanthera

The specific epithet '*platyanthera*' refers to the broad anther. Described by H.G. Reichenbach in 1949 from a plant collected in Semien, northern Ethiopia by Eduard Rueppel.

Plant 20–30 cm tall. Stem erect, leafy. Leaves 3–5, oblong-lanceolate or lanceolate-linear, the longest 6–18 × 0.5–1.8 cm, the upper ones decreasing in size up the stem, the uppermost often bract-like. Inflorescence cylindrical, 5–10 × 1.5 cm, densely many-flowered. Bracts narrowly lanceolate, shorter than the pedicel and ovary. Flowers suberect. Dorsal sepal obovate, convex, 4 × 2 mm. Lateral sepals spreading, very obliquely lanceolate-ovate or lanceolate, 5 × 2.3 mm. Petals erect, curved-elliptical or oblong, 3.6 × 1.2 mm. Lip with a broad undivided base 2 mm long, 3-lobed in the apical three-quarters; lobes ligulate to linear, the midlobe a little longer than the side lobes, 5 × 8 mm; side lobes more or less spreading or recurved, narrower than the midlobe, 4.1 × 0.6 mm; spur inflated and distinctly bilobed in apical half, 8 mm long.

Habitat and distribution In montane grassland and swampy meadows between 1500 and 1600 m in Tigray, Gonder and Wellega. Unknown elsewhere.

Flowering period July to September.

Conservation status Critically endangered.

Notes Differs from *H. epipactidea, H. peristylodes* and *H. zambesiana* by the spur being deeply bilobed at its apex rather than entire. The flowers resemble *Roeperocharis* but the stigmas are not bilobed.

8. H. epipactidea *Rchb.f.*

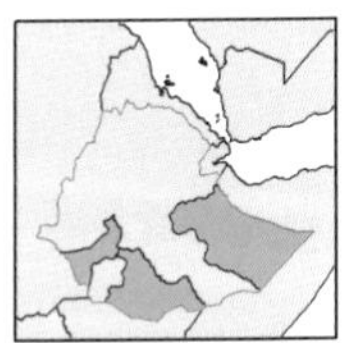

Habenaria epipactidea

The specific epithet '*epipactidea*' refers to the superficial resemblance of the flowers to those of the European terrestrial orchid genus *Epipactis*. Described by H.G. Reichenbach in 1867 from a plant Friedrich Welwitsch collected in Angola.

Plant 15–55 cm tall. Stem erect, usually densely leafy. Leaves 8–15, the lowest 1 or 2 sheath-like, the remainder lanceolate or broadly-lanceolate, the largest 5–12 × 1–2.5 cm, gradually decreasing in size up the stem, the uppermost similar to the bracts. Inflorescence cylindrical, 6–16 × 3–5 cm, densely 7–many-flowered. Bracts lanceolate, shorter than the flowers. Flowers white or greenish-white; ovary with pedicel 15–22 mm long. Dorsal sepal ovate or elliptical, convex, 9–14 × 6–9 mm. Lateral sepals spreading, oblong-lanceolate, longer but narrower than the dorsal. Petals broadly elliptical or almost orbicular, as long as the dorsal sepal, 6–12 mm broad. Lip cuneate at base, 3-lobed almost from the base, midlobe linear, 11–16 × 2 mm; side lobes much shorter and narrower than the midlobe, 4 mm long; spur pendent, swollen towards the apex, 25–31 mm long.

H. epipactidea

Habitat and distribution In open areas with *Acacia* between 1150 and 1800 m in Harerge, Sidamo and Kefa. Also in Kenya, Tanzania, Uganda, Angola, Zimbabwe, Namibia and South Africa.

Flowering period April to July; September.

Conservation status Uncommon and vulnerable.

Notes Differs from *H. zambesina* by its 3-lobed lip and 25–30 mm long spur rather than a more or less entire lip and 30–65 mm long spur.

9. H. peristyloides *A.Rich.*

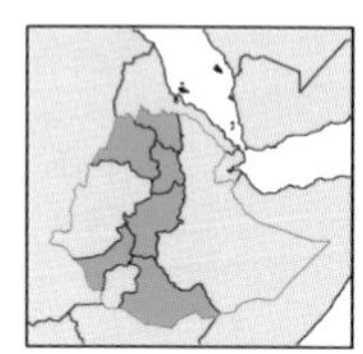
Habenaria peristyloides

The specific epithet '*peristyloides*' refers to the resemblance of the flowers to those of the orchid genus *Peristylus*. Achille Richard described it in 1840 from a plant Richard Quartin-Dillon and Antoine Petit collected in Adwa, Tigray.

Plant 10–80 cm tall. Stem erect, leafy. Leaves 4–9, more or less erect, lanceolate, oblong-lanceolate or lanceolate-linear, the longest 6–24 cm × 0.5–2.5 cm, the upper ones decreasing in size up the stem, the uppermost often bract-like. Inflorescence cylindrical, 5–30 × 1–3 cm, densely many-flowered. Bracts narrowly lanceolate, the lower ones usually longer than the flowers. Flowers green; ovary almost sessile. Dorsal sepal elliptical-lanceolate, convex, 4–6.5 × 2–3 mm. Lateral sepals spreading, very obliquely lanceolate-ovate or lanceolate, 5–9.5 mm long, rather broader than the dorsal. Petals curved-elliptical or oblong, equalling the dorsal sepal. Lip with a broad cordate undivided base, 3-lobed in the apical two-thirds, 7–12 mm long; lobes ligulate to linear, the midlobe 5–9.5 mm long, about twice as long as the side lobes, the side lobes more or less spreading or recurved, narrower than the midlobe; spur cylindrical, 1.5–3 mm long.

Habitat and distribution In short upland grassland, marshes and open scrub between 1650 and 2550 m in Tigray, Gonder, Welo, Shewa, Sidamo and Kefa. Also in Nigeria, Congo, Sudan, Uganda, Kenya and Tanzania.

Flowering period July to September.

Conservation status Local and vulnerable.

Notes Differs from *H. epipactidea* and *H. zambesina* by the spur being 1.5–3 mm long rather than 25 mm long.

10. H. zambesina *Rchb.f.*

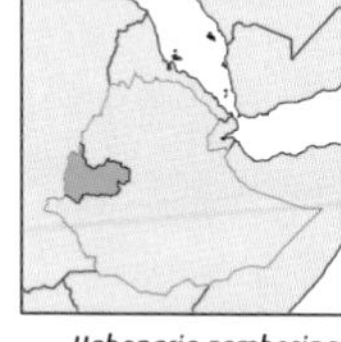
Habenaria zambesina

The specific epithet '*zambesina*' refers to the Zambezi River in Mozambique from where the type specimen was collected by John Kirk. H.G. Reichenbach described the species in 1881.

Plant 40–120 cm tall. Stem erect, stout, leafy. Leaves 9–14, broadly lanceolate or elliptic-ovate, 4–23 × 2–7.5 cm, the uppermost bract-like. Inflorescence cylindrical, up to 40 cm long, densely many-flowered. Bracts lanceolate, the lowermost longer than the flowers.

H. zambesina

Flowers white; pedicel and ovary 15–20 mm long. Dorsal sepal broadly ovate, hooded, 4–6 × 3–5.5 mm. Lateral sepals deflexed, obliquely semi-orbicular, 5–7.5 × 4–6 mm. Petals very obliquely broadly ovate, obtuse, 3.5–6 × 3.5–6 mm. Lip entire or slightly toothed at base, ligulate, acute, convex, 6–9 × 2 mm; spur pendent, narowly cylindrical 3.5–6 cm long.

Habitat and distribution In swampy grassland between 1300–1450 m in Wellega. Also occurs throughout tropical Africa.

Flowering period July and August; October.

Conservation status Rare in Ethiopia but widespread and sometimes common elsewhere.

Notes Differs from *H. epipactidea* by its entire rather than 3-lobed lip and 30–65 mm long spur.

Section **Multipartitae** *Kraenzl.*

A section in which most species are large plants with large flowers with entire petals and a three-lobed lip in which the side lobes are markedly dissected into threads. Two species, *H. excelsa* and *H. taeniodema*, have entire sickle-shaped side lobes to the lip.

Key

1	Side lobes of lip entire	2
-	Side lobes of lip divided into a number of narrow segments	3
2	Dorsal sepal 23–25 mm long; spur 15 mm long; side lobes of lip just exceeding midlobe	**20. H. excelsa**
-	Dorsal sepal 45–50 mm long; spur 20–35 mm long; side lobes of lip just shorter than midlobe	**19. H. taeniodema**
3	Spur at least 10 cm long; sepals and petals 3–4 cm long	**22. H. egregia**
-	Spur up to 4 cm long; sepals and petal less than 2.5 cm long	4

Habenaria lips and spurs

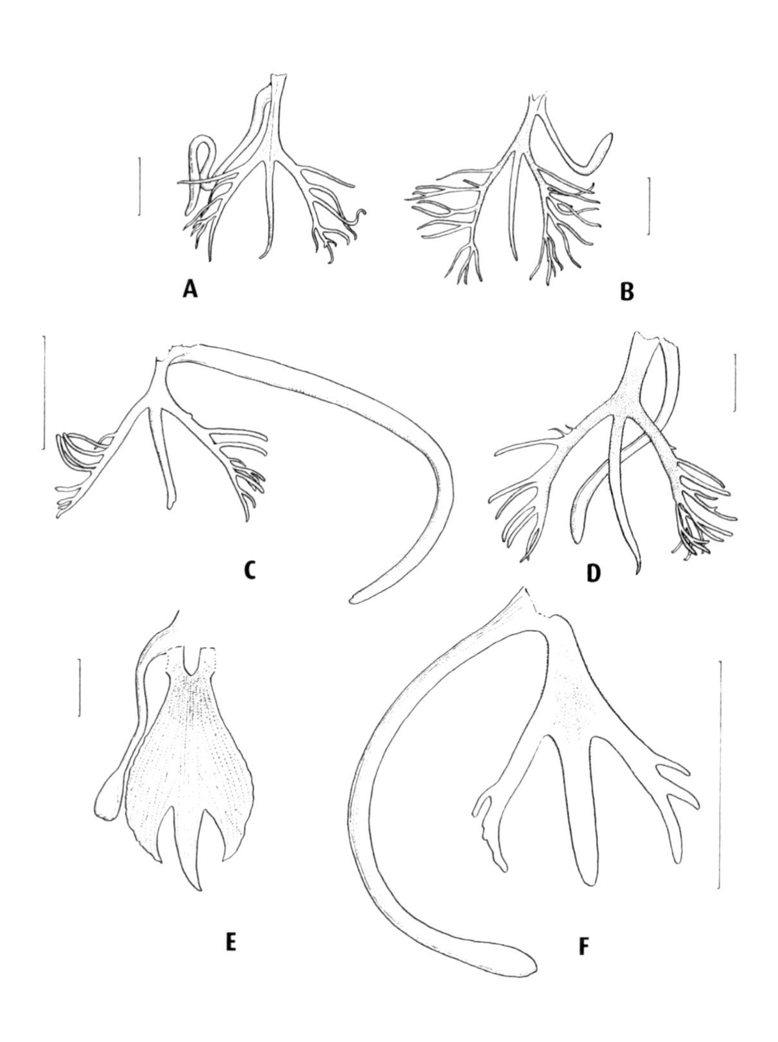

A *H. macrantha*; **B** *H. praestans 5*; **C** *H. quartiniana*; **D** *H. splendens*; **E** *H. taeniodema*; **F** *H. tricruris*. Scale bar = 10mm

Habenaria lips and spurs

A *H. aethiopica*; **B** *H. cavatibrachia*; **C** *H. decorata*; **D** *H. egregia*; **E** *H. excelsa*; **F** *H. gilbertii*. Scale bar = 10mm

4	Anther connective between the two loculi wide (over 8 mm)	5
–	Anther connective between the two loculi narrow (up to 6 mm)	7
5	Basal part of lip glabrous; staminodes not stalked	**17. H. macrantha**
–	Basal part of lip pubescent; staminodes stalked	6
6	Spur 15–25 mm long	**21. H. praestans**
–	Spur 30–40 mm long	**18. H. splendens**
7	Lip undivided for almost half its total length; side lobes of lip not divergent	8
–	Lip undivided for less than one third of its total length; side lobes divergent	10
8	Spur over 11 cm long	**11. H. cavatibrachia**
–	Spur less than 8 cm long	9
9	Spur 2.5–3.5 cm long	**12. H. tricruris**
–	Spur 6.5–8.0 cm long	**13. H. aethiopica**
10	Dorsal sepal only slightly smaller than lateral sepals; midlobe of lip tapering from base	**14. H. decorata**
–	Dorsal sepal much smaller than lateral sepals; midlobe of lip parallel-sided	11
11	Spur sharply folded in lower third; dorsal sepal 10–11 mm long; side lobes of rostellum acute	**16. H. gilbertii**
–	Spur not sharply folded; dorsal sepal 4.5–9 mm long; side lobes of rostellum truncate	**15. H. quartiniana**

11. H. cavatibrachia *Summerh.*

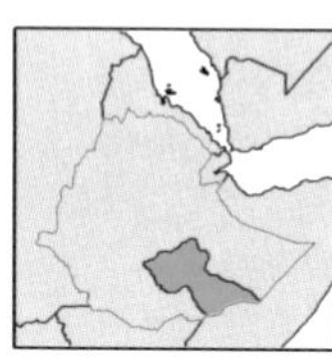

Habenaria cavatibrachia

The specific epithet '*cavatibrachia*' refers to the hollow branches of the stigma. Victor Summerhayes described the species in 1942 from a plant collected south of Mt Elgon in Kenya.

Plant 15–55 cm tall. Stem rather slender, loosely leafy. Leaves 4–7, lanceolate or broadly lanceolate, shortly acuminate, the largest 6–12 × 2–3.5 cm. Inflorescence 5–10 cm long, 2–6-flowered. Bracts shorter than the flowers. Flowers green, the side lobes of the lip whitish, the column brown; pedicel and ovary 3–4 cm long. Dorsal sepal erect, broadly elliptical, convex, 13–20 × 11–13.5 mm. Lateral sepals spreading, obliquely lanceolate, 15–25 × 10 mm. Petals obliquely oblong-elliptical, 11–16 × 6.5–8 mm, ciliate on outer margin. Lip deflexed, 3-lobed just below the middle, wedge-shaped claw 4–9 mm long; midlobe ligulate, 14 mm long; side lobes 16–22 mm long, divided on the outer margin into 3–8 threads which are up to 12 mm long; spur pendent, swollen in the apical half, 11–14 cm long.

Habitat and distribution In montane grassland, between 2100 and 2700 m in Bale. Also in Kenya and Uganda.

Flowering period October.

Conservation status Endangered.

Notes Differs from *H. tricruris* and *H. aethiopica* by the spur being at least 11 cm long, rather than less than 8 cm long.

12. H. tricruris (*A. Rich.*) *Rchb.f.*

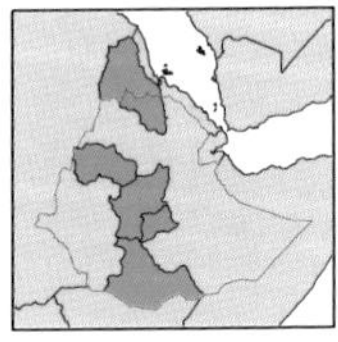

Habenaria tricruris

The specific epithet '*tricruris*' refers to its three-branched lip. Achille Richard described it as *Platanthera tricruris* in 1840 from a plant collected by Richard Quartin-Dillon on Mt Sholloda, near Adwa in Tigray. H.G. Reichenbach transferred it to *Habenaria* in 1855.

Plant 13–30 cm tall. Stem leafy, 2–3 mm diameter. Leaves 4–6, ovate-oblong, largest 40–60 × 20–35 mm. Inflorescence 5–7 cm long, 4–6-flowered. Bracts ovate-lanceolate, up to 25 × 10 mm. Flowers with pale green sepals, petals cream to white, lip pale green; pedicel and ovary 2 cm long. Dorsal sepal erect, ovate, apically ciliate, 10.2–11.5 × 5.5–8.5 mm. Lateral sepals spreading, ovate-oblong, oblique, indistinctly keeled, 12–17 × 6.5–7.5 mm. Petals erect, ovate, ciliate, 9–11 × 5–6 mm. Lip with broad, pubescent, undivided basal part, 5–7 mm long, above this 3-lobed; midlobe linear, 6–8 × 2 mm; side lobes slightly shorter than midlobe, divided almost to the base by 3–7 finger-like fimbriations; spur 25–35 mm, pendent, slightly swollen in apical half.

Habitat and distribution On gentle grassy slopes, in marshy grassland, marshy banks of streams in valley bottoms between 2250 and 2600 m in Tigray, Gojam, Shewa, Arsi and Sidamo and in Eritrea. Unknown elsewhere.

Flowering period June and July.

Conservation status Vulnerable.

Notes Differs from *H. aethiopica* by the spur being 2.5–3.5 cm long, rather than 6.5–8 cm long.

13. H. aethiopica *S.Thomas & P.J.Cribb*

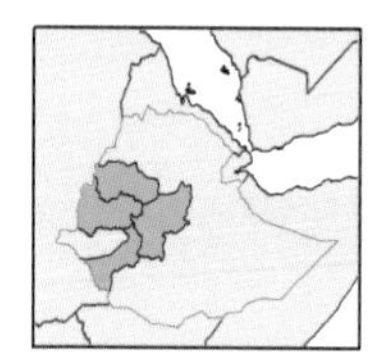

Habenaria aethiopica

The specific epithet '*aethiopica*' refers to the country of origin, Ethiopia, from where the type was collected. Sarah Thomas and Phillip Cribb described it in 1995 from a plant collected in Gojam, north-western Ethiopia.

Plant 30–85 cm tall, stem leafy. Leaves 4–7, ovate-oblong to ovate-lanceolate, largest 70–110 × 30–55 mm, decreasing in size up the stem. Inflorescence 8–13 cm long with 6–13 flowers. Bracts oblong to lanceolate, 8–15 × 20–50 mm. Flowers sweetly scented, sepals green, petals white, lip green fading to white at the base; pedicel and ovary 15–30 mm. Dorsal sepal erect, ovate, apically ciliate, 12.5–16 × 6.5–8.5 mm. Lateral sepals spreading, ovate-oblong, apically keeled, 14–17 × 6–7 mm. Petals erect, oblique, ovate, ciliate, 12.5–16 × 6–7 mm. Lip with broad undivided basal part, pubescent, 7–10 mm long, above this 3-lobed; midlobe linear, rounded, 8–11 × 2 mm; side lobes equalling or just exceeding midlobe, divided almost to the base into 4 or 5 finger-like fimbriations; spur pendent, slightly swollen in apical half, 65–80 mm.

Habitat and distribution On grassy slopes, marshy grassland, amongst *Acacia* bushes, between 1450 and 2450 m in Gojam, Shewa, Kefa and Wellega. Unknown elsewhere.

Flowering period June to August.

Conservation status Endangered.

Notes Differs from *H. tricruris* by the spur being 6.5–8 cm long, rather than 2.5–3.5 cm long.

H. aethiopica

14. H. decorata *A.Rich.*

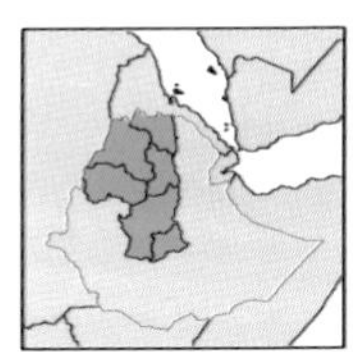

Habenaria decorata

The specific epithet '*decorata*' refers to the beautiful flowers. Achille Richard described it in 1851 from plants collected from Mt Semaiata in Tigray and in the Semien mountains in the Gonder region of northern Ethiopia.

Plant 10–45 cm tall. Stem erect, leafy. Leaves 3–7, lanceolate or oblong-lanceolate, the largest 5–12 × 1.5–4 cm, decreasing in size up the stem. Inflorescence up to 15 cm long, rather laxly 1–11-flowered. Bracts similar to upper leaves, lanceolate, the lower ones overtopping the flowers. Flowers suberect, sepals pale green, petals and lip white; pedicel and ovary 2–2.5 cm long. Dorsal sepal erect, elliptical-ovate, very convex, 8–12 × 5–8 mm. Laterals spreading, obliquely ovate-lanceolate, as long as but narrower than the dorsal. Petals erect, adherent to the dorsal sepal and forming a hood with it, obliquely semi-ovate or almost semi-orbicular, somewhat curved, 8.5–13 × 4.5–8 mm, the outer margins ciliate. Lip with a broad undivided basal part 3–5 mm long, above this 3-lobed; midlobe projecting forwards, narrowly or triangular lanceolate, 10–18 × 3.5–6 mm; side lobes divergent, often curving upwards, 16–22 mm long, divided on the outer margin into 4–6 narrow threads; spur pendent, swollen somewhat in the apical half, 3 cm long.

H. decorata

Habitat and distribution In rocky places in upland moorland with *Erica arborea* and mountain slopes with scattered *Juniperus procera*, between 2500 and 3800 m in Tigray, Welo, Gonder, Gojam, Shewa and Arsi. Also in Kenya and Uganda.

Flowering period July to September.

Conservation status Near threatened.

Notes Differs from *H. gilbertii* and *H. quartiniana* by the midlobe of lip tapering from the base, in contrast to the parallel sided midlobe in both *H. gilbertii* and *H. quartiniana*.

15. H. quartiniana *A.Rich.*

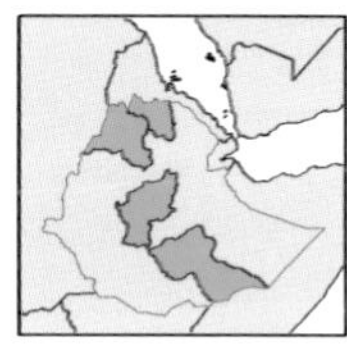
Habenaria quartiniana

The specific epithet '*quartiniana*' was given in honour of the French collector, Richard Quartin-Dillon, who collected the type specimen in 1840 from near Adwa in the Tigray region of northern Ethiopia. Achille Richard described it in 1840.

Plant 25–70 cm tall. Stem erect, leafy. Leaves 5–7, lanceolate or ovate-lanceolate, the largest 6–9.5 × 2–5 cm, decreasing in size up the stem. Inflorescence 7–23 cm long, closely or rather laxly 6–20-flowered. Bracts leaf-like, lanceolate or narrowly lanceolate, the lower ones usually longer than the flowers. Flowers curved outwards, the sepals green, the petals and lip white; pedicel and ovary 2–2.5 cm long. Dorsal sepal erect, lanceolate or elliptical-lanceolate, 4.5–9 mm long. Lateral sepals spreading or deflexed, very obliquely semi-orbicular, curved, 7–13 mm long, broader than the dorsal. Petals obliquely triangular or triangular-lanceolate, somewhat curved, equalling the dorsal sepal but rather broader, ciliate. Lip with a wedge-shaped undivided base 3–4.5 mm long, then deeply 3-lobed; midlobe linear or ligulate, 7–14 mm long; side lobes much diverging, 9–17 mm long, with 6–11 slender threads on the outer margins; spur pendent, somewhat curved, a little swollen in the apical part, 2.5–4 cm long.

Habitat and distribution In upland grassland, light shade amongst shrubs, edges of upland rainforest, often among rocks between 2000 and 2600 mm in Tigray, Gonder, Shewa and Bale. Also in Kenya and Uganda.

Flowering period July and August.

Conservation status Vulnerable.

Notes Differs from *H. gilbertii* in having up to 20 flowers in its inflorescence and by its spur, which is not sharply folded in the lower third.

16. H. gilbertii *S.Thomas & P.J.Cribb*

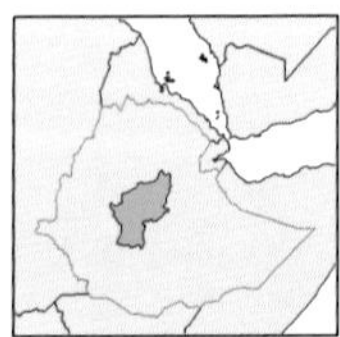
Habenaria gilbertii

The specific epithet '*gilbertii*' was given in honour of the collector of the type specimen, the English botanist Michael Gilbert. Sarah Thomas and Phillip Cribb described it in 1995 by from a specimen collected west of Addis Ababa.

Plant to 45 cm tall. Stem erect, leafy. Leaves 3–5, lowermost sheath-like, largest around centre of stem, 9–11 × 3–3.5 cm, oblong, acute. Inflorescence 14–16 × 6–7 cm, 9–11-flowered, lax. Bracts lanceolate, decreasing in size towards apex of inflorescence. Flowers pale green; pedicel and ovary 2 cm long. Dorsal sepal erect, elliptic, ciliate, 10.6–11.5 × 6.1–7.6 mm. Lateral sepals spreading, ovate-oblong, sub-oblique, ciliate, mid-nerve keeled, 12–14 × 5–5.2 mm. Petals oblong, ciliate, 11–11.5 × 4.5–4.7 mm. Lip with undivided basal part 5 mm long then deeply 3-lobed; midlobe linear, pubescent, 10 × 1 mm, side lobes diverging, 14–17 mm long, 6–12 slender threads along outer margin, longest 7 mm; spur hook-like, 37–50 mm, pendent, strongly reflexed through 180° in apical third, and lightly reflexed just above this.

Habitat and distribution In humus, grassy spots among cultivated fields and in wet evergreen bushland with *Carissa edulis* between 2100 and 2300 m in Shewa. Unknown elsewhere.

Flowering period June and July.

Conservation status Critically endangered.

Notes Differs from *H. quartiniana* by the spur being sharply folded (hook-shaped) in the lower third and in having a 9–11-flowered inflorescence.

17. H. macrantha *A.Rich.*

The specific epithet '*macrantha*' refers to the large flowers. Achille Richard described it in 1851 from plants collected from the Semien mountains in Gonder.

Habenaria gilbertii

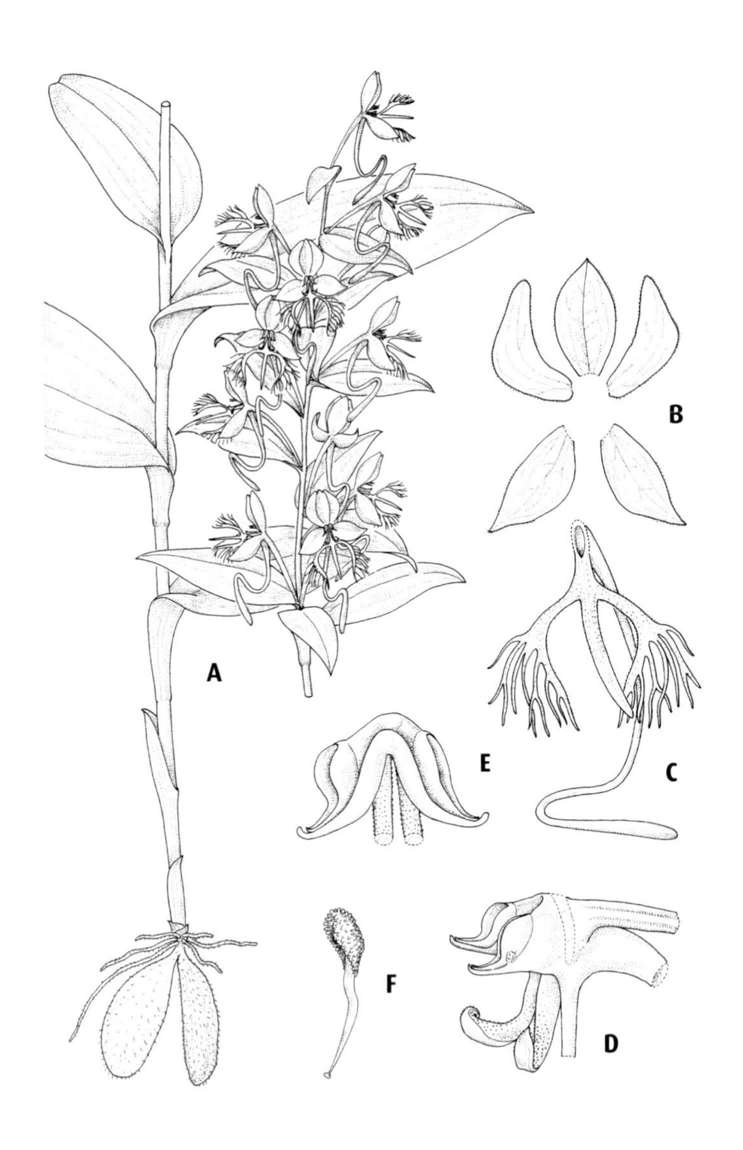

A habit × ½; **B** sepals and petals × 2; **C** lip and spur × 1½; **D** column, base of lip and ovary, side view × 4; **E** column, front view × 4; **F** pollinium × 4. All drawn from *Gilbert* 2149 except E from *de Wilde* 5176 by Susanna Stuart-Smith.

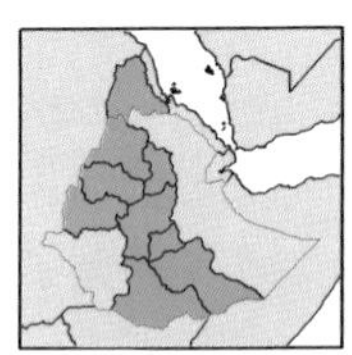
Habenaria macrantha

Plant 20–50 cm tall. Stem erect, leafy. Leaves 5–7, lanceolate, ovate-lanceolate or elliptical-lanceolate, the largest 5–12 × 1.5–5 cm, the upper ones smaller, similar to the bracts. Inflorescence up to 12 cm long, rather loosely 2–9-flowered. Bracts leaf-like, lanceolate, the lower ones equalling or longer than the flowers. Flowers suberect, green or whitish-green; pedicel and ovary 2.3–3 cm long. Dorsal sepal erect, adherent to petals, ovate to narrowly lanceolate, 20–26 × 7–11 mm. Lateral sepals spreading or deflexed, obliquely lanceolate; all sepals ciliate. Petals adherent to the dorsal sepal, lanceolate, 20–25 × 7–11 mm, very shortly pubescent-hairy and ciliate. Lip united to the base of the column for 6 mm, almost glabrous, with a narrow undivided claw 9–15 mm long, then 3-lobed, 3 cm long; midlobe linear, 14–23 × 1–2 mm, with a longitudinal ridge; side lobes diverging but more or less incurved, rather longer than the midlobe, with 6–10 narrow threads on the outer margins; spur more or less incurved, often sharply bent in the middle, slightly swollen in the apical half, 20–35 mm long.

Habitat and distribution In upland grassland and upland moor among *Erica arborea* and *Juniperus procera* between 1650 and 3200 m in Welo, Gonder, Gojam, Shewa, Arsi, Bale, Sidamo and Wellega and in Eritrea. Also in Uganda, Kenya, Somalia and Yemen.

Flowering period July to September.

Conservation status Near threatened.

Notes Differs from *H. praestans* and *H. splendens* by the basal part of the lip being glabrous rather than pubescent.

18. H. splendens *Rendle*

The specific epithet '*splendens*' refers to the fine large flowers. Described by Alfred Rendle in 1895 from plants collected in Tanzania.

Plant 30–75 cm tall. Stem erect, leafy. Leaves 6–8, lanceolate, ovate-lanceolate or ovate, the largest 6–20 × 2.5–8 cm. Inflorescence 8–27 × 6–10 cm, rather laxly 4–17-flowered. Bracts lanceolate or broadly lanceolate, the lowest ones equalling or longer than the flowers. Flowers suberect, sepals pale green, petals and lip white, often fragrant; pedicel and ovary 2.5–3 cm long. Dorsal sepal erect, broadly lanceolate or lanceolate-

H. splendens

elliptical, 20–30 × 9–15 mm. Lateral sepals spreading, very obliquely lanceolate, a little longer than the dorsal but narrower. Petals erect, adherent to the dorsal sepal, very curved, ligulate or lanceolate-ligulate, 20–30 × 5–9 mm. Lip with a narrow undivided claw 7–13 mm long, then deeply 3-lobed, the claw and lobe-bases densely pubescent; midlobe often more or less deflexed, linear, 2–3 × 1–2 mm; side lobes diverging, a little longer than the midlobe with 6–12 threads on the outer margins; spur incurved, S-shaped, swollen in apical part, 3–4 cm long.

Habitat and distribution Said to occur in Ethiopia by Summerhayes (1968) in the *Flora of Tropical East Africa*, but so far no Ethiopian specimens have been collected. It occurs in Kenya, Tanzania, Uganda, Malawi, Zambia.

Flowering period December to February in East Africa.

Conservation status If present then very rare in Ethiopia.

Notes Differs from *H. praestans* by the spur being 3–4 cm long rather than 1.5–2.5 cm long.

19. H. taeniodema *Summerh.*

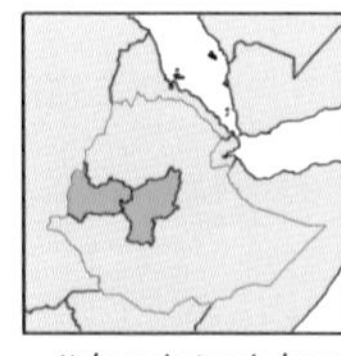
Habenaria taeniodema

The specific epithet '*taeniodema*' refers to the band-like or strap-shaped connective. Victor Summerhayes described it in 1966 from a plant collected by Kurt Hildebrandt in Shewa, central Ethiopia.

Plant to 1 m tall. Leaves suberect, 3–4, narrowly-lanceolate, largest ones 12–19 × 2.5–4 cm. Inflorescence up to 20 cm long, up to 5-flowered. Bracts lanceolate, 8–12 × 1.5–3 cm. Flowers 8 cm in diameter; pedicel and ovary 2.5–3.5 cm long. Dorsal sepal convex, lanceolate, 45–50 × 25–30 cm. Lateral sepals narrowly lanceolate, 45–50 × 7 mm, adnate to column for 6 mm at the base. Petals erect, oblanceolate, shortly acuminate, 50–55 × 15 mm, adnate to column for 5 mm at the base. Lip porrect, obovate-oblong, 50–55 × 20–25 mm, 3-lobed at apex; midlobe linear-triangular, 15–20 × 5 mm; side lobes just shorter than midlobe, triangular, 14 × 7 mm; spur 20–35 × 5–7 mm, apex inflated.

Habitat and distribution In damp scrub between 2200 and 2400 m in Shewa and Wellega. Unknown elsewhere.

Flowering period August and September.

Conservation status Critically endangered.

Notes Differs from *H. excelsa* in having large flowers 8 cm across, with a 4.5–5 cm long dorsal sepal rather than smaller flowers with a 2.3–2.5 cm long dorsal sepal.

20. H. excelsa *S.Thomas & P.J.Cribb*

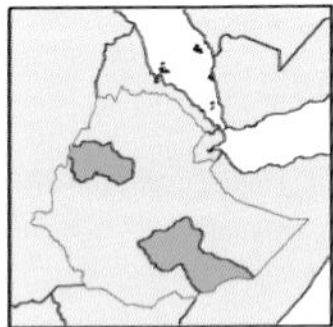
Habenaria excelsa

The specific epithet '*excelsa*' refers to the fine flowers. It was described by Sarah Thomas and Phillip Cribb in 1995 from plants collected in the Gojam region in north-western Ethiopia.

Plant to 37 cm tall. Stem 4–5 mm diameter, leafy. Leaves 5–7, linear lanceolate, largest 13–15 × 1.7–2 cm. Inflorescence 5–10 × 5 cm, 3–5-flowered. Bracts ovate-oblong, 3–5 × 1–1.5 cm. Flowers green with yellow-orange tint; pedicel and ovary 18–22 mm long. Dorsal sepal concave, obovate, 23–25 × 10–11 mm. Lateral sepals ovate-lanceolate, oblique, shortly cuspidate, 20–23 × 6–8 mm, fused to column for 3–4 mm at base. Petals oblique, obovate, minutely ciliate, 23–25 × 7.5–8 mm, fused to column for 3–4 mm at base. Lip 18–21 mm long, undivided basal part 9–11 × 10–13 mm, fused to column for 4 mm, 3-lobed above; midlobe 7–8.5 mm, linear to narrowly triangular; side lobes just exceeding midlobe, spreading, narrowly falcate, curving away from midlobe; spur 15 × 2–3 mm, swollen at base.

Habitat and distribution In rock crevices between 3150 and 3500 m in Gojam and Bale. Unknown elsewhere.

Flowering period July and August.

Conservation status Critically endangered.

Notes Differs from *H. taeniodema* in having smaller flowers with a 2.3–2.5 cm long dorsal sepal.

Habenaria excelsa

A habit × ½; **B** flower, side view × 1; **C** sepals, petals and lip flattened × ⅔; **D** column from above × 3; **E** column and base of lateral sepal from side × 3; **F** rostellum × 3; **G** pollinium × 4. A drawn from the type collection; B – G drawn from *Mesfin* T5123. All drawn by Susanna Stuart-Smith.

21. H. praestans *Rendle*

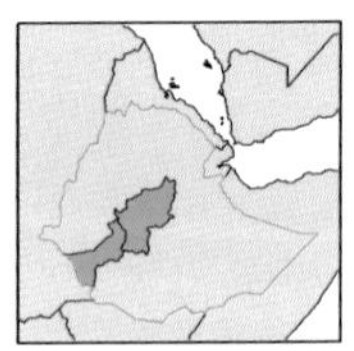
Habenaria praestans

The specific epithet '*praestans*' refers to the fine "pre-eminent" flowers. Alfred Rendle described it in 1895 from plants collected from the Ruwenzori Mountains in Uganda.

Robust plant 25–105 cm tall. Stem erect, leafy, often stout and up to 1 cm in diameter at the base. Leaves 6–12, lanceolate or broadly lanceolate, the largest 8–30 × 2.5–7.5 cm, decreasing in size up the stem, the uppermost similar to the bracts. Inflorescence cylindrical, 10–30 cm long, densely 4–30-flowered, 5–10 cm in diameter. Bracts leaf-like, lanceolate, the lower ones usually longer than the flowers. Flowers suberect, green with white lip; pedicel and ovary 2–2.5 cm long. Dorsal sepal erect, broadly lanceolate or elliptical-lanceolate, 1.5–2.7 × 1–1.5 cm. Lateral sepals spreading, obliquely broadly lanceolate, equal in size to the dorsal. Petals adherent to the dorsal sepal, much curved, ligulate-lanceolate or ligulate. Lip with a linear undivided part 7–10 mm long, then deeply 3-lobed, 3–4 cm long, thickly pubescent on the claw and main lobes; midlobe linear, side lobes diverging, somewhat longer than the midlobe with 8–15 slender threads on the outer margins; spur slightly curved, a little swollen towards the end, 1.5–2.5 cm long.

Habitat and distribution In upland grassland and in shade of shrubs between 1250 and 2400 m in Shewa and Kefa. Also in Kenya, Tanzania, Uganda, Malawi, Zimbabwe, Zambia, Rwanda.

Flowering period August.

Conservation status Endangered.

Notes Differs from *H. splendens* by the spur being 1.5–2.5 cm long rather than 3–4 cm long.

H. praestans

22. H. egregia *Summerh.*

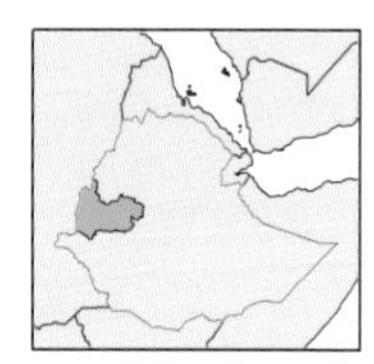

Habenaria egregia

The specific epithet comes from the Latin *egregius*, meaning distinguished, in allusion to its magnificent flowers. Victor Summerhayes described it in 1949 from a specimen collected in Cameroon.

Plant large, glabrous, 35–50 cm tall. Stems stout, 5 mm in diameter, leafy. Leaves suberect-spreading, ovate, 6.5–8 × 4.5–5.2 cm, sheathing at base, drying blackish brown. Inflorescence 2- or 3-flowered; bracts lanceolate, 3.5–6.3 × up to 2.6 cm. Flowers very large, showy, sweetly scented, sepals green, petals and lip white; pedicel and ovary 2–3.7 cm long. Dorsal sepal oblong-elliptic, 2.7–3 ×1.1–1.3 cm, forming a hood with the petals. Lateral sepals obliquely lanceolate, 3.2–3.7 × 0.9–1.1 cm. Petals obliquely oblong-lanceolate, 3.2–3.5 × 1–1.3 cm. Lip 3-lobed, 4.2–6 cm long, with a slender basal claw 1–2 cm long; side lobes spreading, 2–2.8 cm long, deeply lacerate-fimbriate; midlobe linear, slightly longer than the side lobes; spur sometimes slightly S-shaped, cylindrical, 9–13 cm long.

H. egregia

Habitat and distribution In shade under *Acanthus* in bushy meadows with rocky outcrops, bamboo thicket between 1500 and 1600 m. Known only in Benishangul-Gumuz Province in Wellega. Also in Cameroon and two collections from Kavirondo District in Kenya.

Conservation status Endangered throughout its range. Known in Ethiopia from only two specimens.

Flowering period June and July.

Notes This is the largest flowered species of *Habenaria* in Ethiopia and a magnificent plant. It differs from all other species in sect. *Multipartitae* in Ethiopia in having two or three large flowers with a 3–4 cm long dorsal sepal, a long spur (usually over 10 cm) and an anther-connective over 10 mm wide.

Section **Pentaceras** (*Thou.*) *Schltr.*

A section in which most species have leaves borne along the stem, divided petals, a spur less than 3 cm long, and an erect dorsal sepal.

A single species has been reported from Ethiopia.

23. H. malacophylla *Rchb.f.*

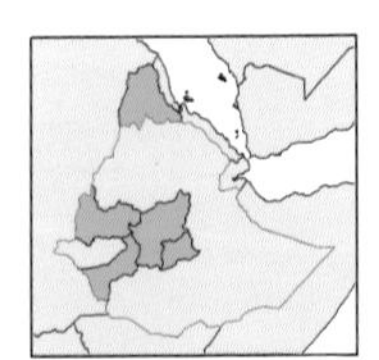
Habenaria malacophylla

The specific epithet '*malacophylla*' refers to the soft leaves. H.G. Reichenbach described it in 1881 from a plant collected from the Cape Province in South Africa.

Plant 30 cm to 1 m tall, glabrous. Stem leafy in centre with a rather bare lower part. Leaves 10–19, oblanceolate, with a rather narrow basal part above the sheath, acute, the largest 7–20 × 2–5 cm, the upper 2 or 3 smaller, lanceolate. Inflorescence 8–34 × 2.5–3.5 cm, many-flowered. Bracts lanceolate, 9–20 mm long, shorter than the flowers. Flowers green; pedicel and ovary 10–17 mm long, bow-shaped. Dorsal sepal erect, broadly elliptical ovate, convex, 3.6–3 × 2.5–4.5 mm broad. Lateral sepals deflexed, obliquely lanceolate, 4–7.5 × 2–3.3 mm. Petals bipartite nearly to the base; posterior lobe adherent to dorsal sepal, ligulate, 4–7.3 × 0.5–1.5 mm, curved; anterior lobe curved upwards from near base, linear, 4.5–9.5 mm long, narrower than the posterior. Lip projecting outwards, 3-lobed nearly to the base; lobes linear;

H. malacophylla

midlobe 4.5–7 mm long, the side lobes usually a little longer; spur pendent, somewhat thicker in the middle than at either end, 9–18 mm long.

Habitat and distribution In upland rainforest but rarely in upland grassland in forested areas between 1900 and 2400 m in Shewa, Arsi, Kefa and Wellega and in Eritrea. Also in Sierra Leone to Nigeria, DR Congo, Uganda, Kenya, Tanzania, Malawi, Zambia, Zimbabwe and South Africa (Transvaal, Natal, Cape Province), and Oman.

Flowering period May; August and September; November and December.

Conservation status Vulnerable throughout its range.

Section **Replicatae** *Kraenzl.*

A section in which most species have cauline leaves and flowers that resembles gnats or mosquitoes, with divided petals and a reflexed dorsal sepal.

Five species have been reported from Ethiopia.

Key

1	Spur strongly spirally twisted at least once in the middle	**25. H. schimperiana**
–	Spur not spirally twisted or scarcely so	2
2	Ovary less than one quarter the length of the pedicel	**27. H. ichneumonea**
–	Ovary more than one third the length of the pedicel	3
3	Spur narrow-cylindrical, not inflated at or near apex	**24. H. chirensis**
–	Spur inflated at the apex	4
4	Side lobes of lip shorter than midlobe; stigmas 3–5 mm long, spur 15–25 mm long	**26. H. humilior**
–	Side lobes of lip longer than midlobe, stigmas 2 mm long, spur 10–13 mm long	**28. H. vollesenii**

24. H. chirensis *Rchb.f.*

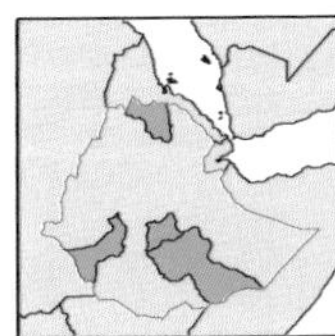

Habenaria chirensis

The specific epithet '*chirensis*' refers to Chire, in Tigray where the type was collected by Richard Quartin-Dillon. H.G. Reichenbach described it in 1881.

Plant 20 cm to 1 m tall, glabrous. Stem erect, leafy throughout its length. Leaves 7–13, linear or lanceolate-linear, the largest 8–30 × 0.5–1.5 cm, the upper ones smaller. Inflorescence 6–30 × 3–5 cm, 12- to many-flowered. Bracts lanceolate, distinctly shorter than the pedicel with ovary. Flowers white or greenish-white, with an unpleasant smell; pedicel and ovary 2 cm long. Dorsal sepal narrowly elliptical, convex, 4.5–6 × 2 mm. Lateral sepals obliquely obovate, 6–9 × 3–5 mm. Petals bipartite nearly to the base, papillose and ciliate; posterior (upper) lobe reflexed, linear, 4–6 × 0.5 mm; anterior lobe spreading forwards, narrowly lanceolate, 5–9 mm × 1 mm. Lip tripartite from a short undivided part; lobes linear, the midlobe 9–12 × 0.5 mm, the side lobes 6–8 mm long, narrower than the midlobe; spur parallel to the ovary or more or less incurved, equally cylindrical almost to the apex which is very slightly swollen and truncate, 1–2 cm long.

Habitat and distribution In damp grassland, swamps, or wet places among rocks between 1150 and 2600 m in Tigray, Arsi, Bale and Kefa. Also in Nigeria, Cameroon, Uganda, Kenya and Tanzania.

Flowering period June to September.

Habenaria petals, lips and spurs

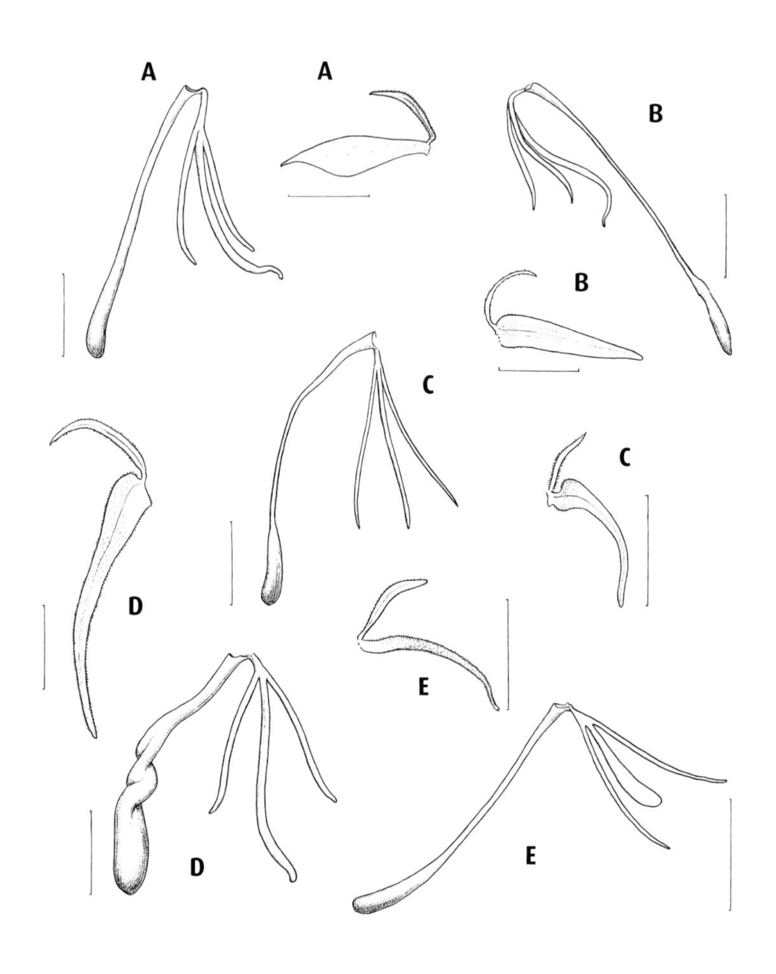

A *H. chirensis;* **B** *H. humilior;* **C** *H. ichneumonea;* **D** *H. schimperiana;*
E *H. vollesenii.* Scale bar = 5 mm.

Conservation status Locally common.

Notes Differs from *H. humilior* and *H. vollesenii* by the spur being narrow-cylindrical rather than inflated at or near apex.

25. H. schimperiana *A.Rich.*

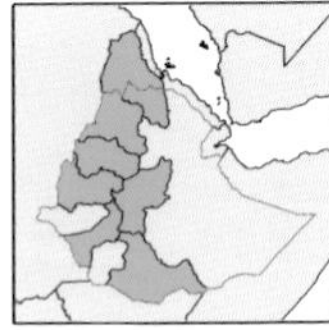

Habenaria schimperiana

The specific epithet '*schimperiana*' was given in honour of Georg Wilhelm Schimper, who collected the type specimen from Semien, Gonder. Achille Richard described it in 1851.

Plant 30 cm to 1 m tall, glabrous. Stem erect, leafy throughout its length. Leaves 6–10, linear or linear-lanceolate, the largest 7–28 × 1–2 (rarely –3–7) cm, the upper ones smaller. Inflorescence 6–35 × 5.5–10 cm, rather loosely or sometimes more densely 4- to many-flowered. Bracts lanceolate, usually much shorter than the pedicel with ovary. Flowers green with white central parts, with an unpleasant smell; pedicel with ovary 2–3.5 cm long. Dorsal sepal reflexed, narrowly elliptical, convex, 6–8 × 4 mm. Lateral sepals deflexed and twisted, obliquely obovate with apiculum lateral,

H. schimperiana

H. humilior

9–11 × 5–8 mm. Petals bipartite nearly to the base, both lobes ciliate, especially the anterior; posterior lobe reflexed, linear or narrowly lanceolate, 5–8 × 1 mm; anterior lobe spreading downwards, much longer, elongate-lanceolate, 14–18.5 × 2 mm. Lip deflexed or incurved, tripartite from an undivided base 2–3 mm long; midlobe linear, 13–17 mm long; side lobes very narrowly lanceolate, 8–11 mm long; all lobes 0.5 mm broad; spur several times twisted in the middle, much swollen in the apical half, 10–16 mm long.

Habitat and distribution In swamps or wet grassland on badly drained soil between 1450 and 2550 m in Tigray, Gonder, Gojam, Shewa, Sidamo, Kefa and Wellega and in Eritrea. Also in Sudan, DR Congo, Kenya, Tanzania, Malawi, Zambia, Zimbabwe and Yemen.

Flowering period June in Sidamo; July to October elsewhere in Ethiopia.

Conservation status Locally common.

Notes Distinguished from related species by the spur being strongly spirally twisted.

26. H. humilior *Rchb.f.*

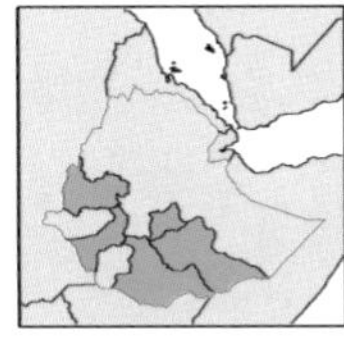

Habenaria humilior

The specific epithet '*humilior*' refers to the insignificant flowers. H.G. Reichenbach described it in 1881 from a plant collected from Gonder by Georg Wilhelm Schimper.

Plant 15–70 cm tall. Stem leafy throughout its length. Leaves 7–13, lanceolate or almost linear, the largest 6–23 × 0.5–3 cm, the upper ones much smaller, similar to the lower bracts. Inflorescence 5–25 × 3.5–5 cm, rather laxly to densely 6- to many-flowered. Bracts lanceolate, 1–2 cm long, usually somewhat shorter than the pedicel and ovary. Flowers green or greenish-white; pedicel and ovary 1.5–2 cm long. Dorsal sepal reflexed, narrowly elliptical, convex, 4–6.5 × 1.5–3 mm. Lateral sepals deflexed, obliquely obovate, 6–9.5 × 3–5 mm. Petals bipartite nearly to the base; posterior (upper) lobe more or less reflexed, linear, 4–6 × 0.5 mm, ciliate; anterior lobe much longer and broader, elliptical-ligulate, narrowly oblong or oblong-lanceolate, 6–11 × 1.5–3 mm. Lip deflexed, tripartite from an undivided base less than 1 mm long; midlobe linear, incurved, 6–11.5 × 0.5–1 mm; side lobes lanceolate-linear, 5–9 × 0.1–1 mm; spur only slightly twisted, swollen in the apical half, 1.5–2.5 cm long.

Habitat and distribution In short grassland, often on shallow poorly drained soil over rocks between 1600 and 2400 m in Gonder, Shewa, Arsi, Bale, Sidamo, Kefa and Wellega and in Eritrea. Also in Congo, DR Congo, Sudan, Uganda, Kenya, Tanzania, Malawi, Zambia and Zimbabwe.

Flowering period May and June in Sidamo; July to September elsewhere in Ethiopia.

Conservation status Locally common.

Notes Differs from *H. vollesenii* by the spur being 15–25 mm long rather than 10–13 mm long.

27. H. ichneumonea (*Sw.*) *Lindl.*

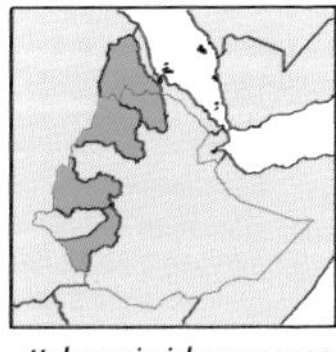

Habenaria ichneumonea

The specific epithet '*ichneumonea*' refers to the ichneumon fly-like flowers. Described as *Orchis ichneumonea* by Olof Swartz in 1805 from a plant collected by Afzelius in Sierra Leone.

Plant 20–85 cm tall. Stem leafy throughout its length. Leaves 5–12, linear, the largest 7–20 × 5–10 mm, the upper ones smaller. Inflorescence 6–22 × 3–5 cm, rather loosely 7–many-flowered. Bracts lanceolate, usually much shorter than the pedicel and ovary. Flowers green with white centre; pedicel and ovary 1.5–3 cm long. Dorsal sepal reflexed, narrowly elliptical, convex, 3.5–5 × 1.5–2.5 mm. Lateral sepals deflexed, obliquely elliptical or obovate, 5–8.5 × 3–5 mm. Petals bipartite nearly to the base; posterior (upper) lobe reflexed, linear, 3–4.5 × 0–5 mm, ciliate; anterior lobe spreading forwards, narrowly lanceolate, 6.5–11 × 1.5 mm, glabrous. Lip deflexed, tripartite from an undivided base, 1.5–4 mm long; lobes linear, more or less incurved, the midlobe 7–13 mm long, the side lobes usually a little shorter, all less than 0.5 mm broad; spur 10–25 mm long with apex much swollen.

Habitat and distribution In damp grasslands or swamps between 1000 and 1300 m in Tigray, Gonder and Kefa and in Eritrea. Also occurs from Guinea and Senegal to DR Congo, Burundi, Uganda, Tanzania, Angola, Malawi, Zambia, Zimbabwe and Botswana.

Flowering period July and August.

Conservation status Locally common.

Notes Differs from related species by the ovary being less than a third the length of the pedicel.

28. H. vollesenii *S.Thomas & P.J.Cribb*

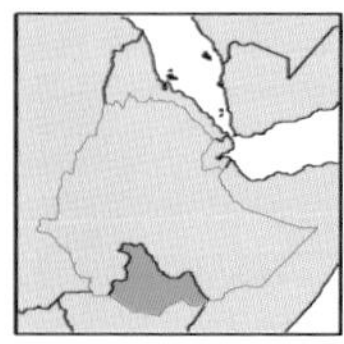

Habenaria vollesenii

The specific epithet '*vollesenii*' was given in honour of the Danish botanist, Kaj Vollesen, one of the collectors of the type specimen. Described by Sarah Thomas and Phillip Cribb in 1995.

Plant 25–40 cm tall, glabrous. Stem leafy in lower half. Leaves 2–5, lanceolate or almost linear, the largest towards the base, 50–110 × 8–14 mm, the upper ones much smaller, similar to bracts. Inflorescence 12–14 × 3–4 cm, rather laxly 20–35-flowered. Bracts narrowly lanceolate, usually somewhat shorter than the pedicel with ovary. Flowers green or greenish-white; pedicel with ovary straight or more or less curved, 10–15 mm long. Dorsal sepal more or less reflexed, elliptic, convex, 3.2–3.9 × 1.7–2.2 mm. Lateral sepals deflexed, obliquely obovate, 4–5.2 × 2.3–2.8 mm. Petals bipartite nearly to the base; posterior (upper) lobe more or less erect, linear, 3.5–4 × 0.5 mm, ciliate; anterior lobe much longer, linear, ciliate, 7.5–8.7 × 0.5–0.3 mm. Lip tripartite from an undivided base less than 1 mm long; midlobe linear, rounded, 5.5–6.5 × 0.6–1.2 mm; side lobes lanceolate-linear, 7.5–8 × 0.4–0.5 mm; spur 10–13 cm long, slightly swollen in the apical half.

Habitat and distribution In low woodland/high bushland and with *Acacia* and *Lannea rivae* between 1200 and 1575 m in Sidamo. Unknown elsewhere.

Flowering period April to June.

Conservation status Critically endangered.

Notes Differs from *H. humilior* by the spur being 10–13 mm long rather than 15–25 mm long.

H. vollesenii

Habenaria vollesenii

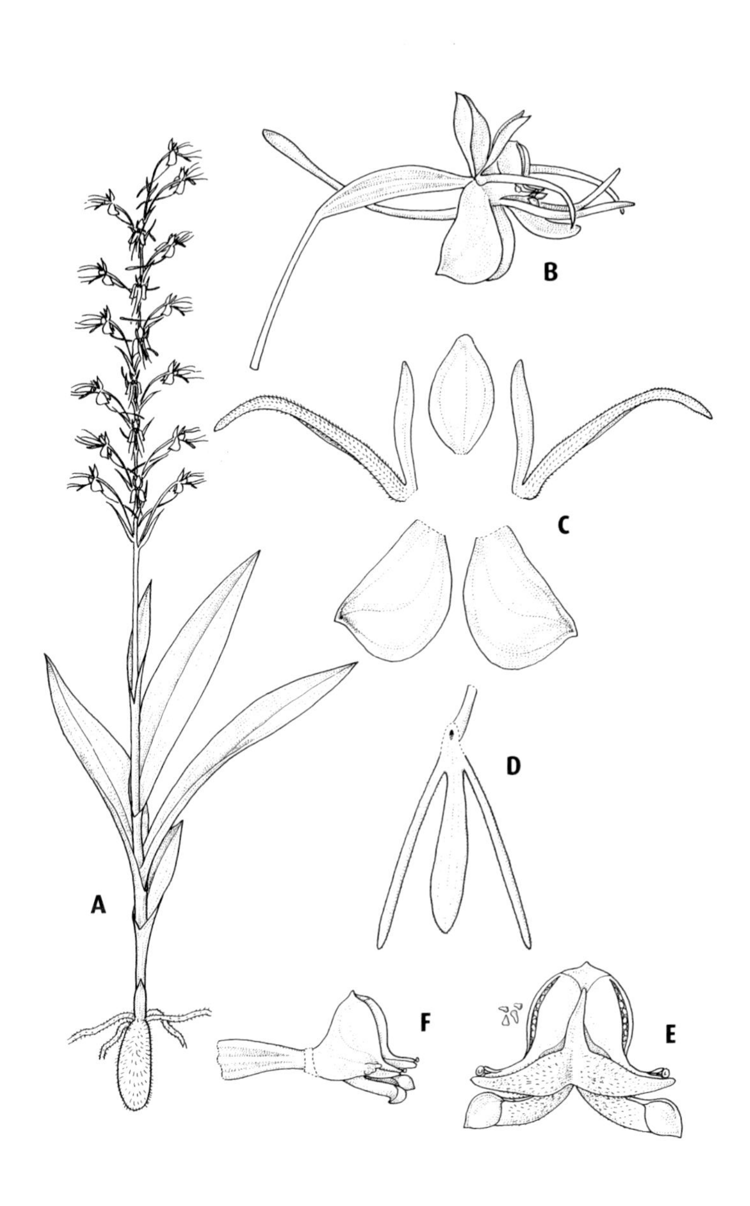

A habit × ½; **B** flower, side view × 3; **C** sepals and petals flattened × 4; **D** lip flattened × 4; **E** column from front × 7; **F** column, side view × 4. All drawn from *Mesfin & Vollesen* 4139 by Susanna Stuart-Smith.

Section **Cultratae** *Kraenzl.*

A section in which most species somewhat resemble those of sect. *Replicatae* but have an erect rather than reflexed dorsal sepal. Abnormal, malformed flowers are common in this section.

Five somewhat ill-defined species have been reported from Ethiopia.

Key

1	Lip entire; bracts overtopping flowers	**32. H. rivae**
–	Lip 3-lobed; bracts not overtopping flowers	2
2	Spur less than 15 mm long	3
–	Spur more than 17 mm long	4
3	Plants more than 40 cm tall; leaves all along stem, largest up to 19 × 4.8 cm; stigma lobes 5 mm long, clavate	**30. H. cultrata**
–	Plants usually less than 30 cm tall; leaves towards base of stem, largest less than 8 × 2.7 cm; stigmas 2–4 mm long	**29. H. antennifera**
4	Spur usually 28 mm or more long; stigma lobes less than 2.5 mm long; lateral sepals with apex slightly eccentrically placed at top; inflorescences usually 5–6 cm in diameter	**33. H. tweedieae**
–	Spur 17–23 mm long; stigma lobes 3 mm or more long; lateral sepals with a very eccentrically placed apex; inflorescence usually 4–5 cm in diameter	**31. H. cultriformis**

29. H. antennifera *A.Rich.*

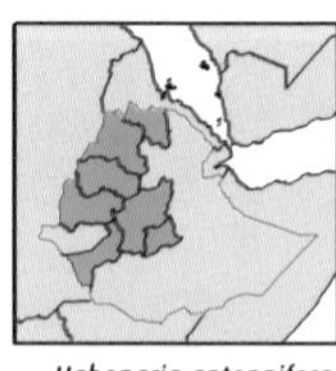
Habenaria antennifera

The specific epithet '*antennifera*', bearing antennae, refers to the antennae-like segments of the petals and lip. Achille Richard described it in 1840 from a plant collected by Quartin-Dillon near Adwa, Tigray.

Plant up to 45 cm tall. Stems leafy throughout length. Leaves 4–7, ovate to elliptic-lanceolate, the largest 4–14 × 2–3.5 cm. Inflorescence 4.5–20 × 2.5–3.5 cm diam., cylindrical, subdensely many-flowered. Bracts mostly longer than the ovary, ovate-elliptic, glandular-pubescent. Flowers green, yellow-green or green and white. Dorsal sepal 3.7–4.2 × 1.7–2.5 mm, ovate to elliptic, concave, lying over the column; lateral sepals 6.5–7 × 5–5.5 mm, reflexed, obliquely oblong-elliptic. Petals bipartite to base; upper (posterior) lobe 3.5–4 × 1–1.5 mm, oblanceolate, ciliate; lower (anterior) lobe up to 12 × 2 mm, reflexed, linear-

H. antennifera

H. cultrata

lanceolate. Lip 3-lobed at base; side lobes 5–7 × 1 mm, linear to linear-lanceolate; midlobe 6–9 × 1–1.7 mm, linear to lanceolate; spur 9–12 mm long, clavate, straight or slightly geniculate in middle but never very twisted.

Habitat and distribution In grassland or open scrub, often amongst rock between 2000 and 3300 m in Tigray, Gonder, Welo, Gojam, Shewa, Arsi and Kefa. Also in Yemen.

Flowering period August to October.

Conservation status Locally common but vulnerable.

Notes Differs from *H. cultrata* by the plant being less than 30 cm tall rather than more than 40 cm tall.

30. H. cultrata *A.Rich.*

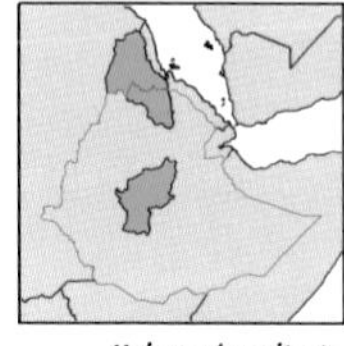

Habenaria cultrata

The specific epithet '*cultrata*' refers to the knife-shaped leaves. Achille Richard described it in 1851 from a plant collected by Richard Quartin-Dillon in Chire, Tigray.

Plant 30–60 cm tall. Stem leafy throughout its length. Leaves 5–9, the largest in the middle of the stem, 9–24 × 2–4 cm, narrowly elliptic or elliptic-lanceolate. Inflorescence 10–16 cm long, laxly many-flowered. Bracts 1–1.7 cm, shorter than the pedicel with ovary, lanceolate to oblanceolate. Flowers greenish, sometimes apparently cleistogamous. Dorsal sepal 3.5–4.5 × 1.5–2.6 mm, elliptic, concave, lying over the column; lateral sepals 6–8 × 4–6 mm, reflexed, with an obliquely set blunt apex, slightly pubescent within. Petals bipartite at the base; upper (posterior) lobe 3.5–4 × 0.8 mm, more or less adnate to dorsal sepal, linear, ciliate; lower (anterior) lobe 7.5–11.5 × 2–2.6 mm, deflexed, lanceolate, pubescent. Lip shortly clawed, 3-lobed almost at base; side lobes 5–7 mm long, linear; midlobe 9–14 mm long, linear; spur 12–15 mm long, dilated towards apex.

Habitat and distribution In grassland near permanent water or in shade of bushes or amongst rocks betwen 1700 and 2100 m in Tigray and Shewa and in Eritrea. Also in Yemen and Oman.

Flowering period July to September.

Conservation status Locally common but vulnerable.

Notes Differs from *H. antennifera* by the plant being more than 40 cm tall.

H. cultrata abnormal form

31. H. cultriformis *Kraenzl.*

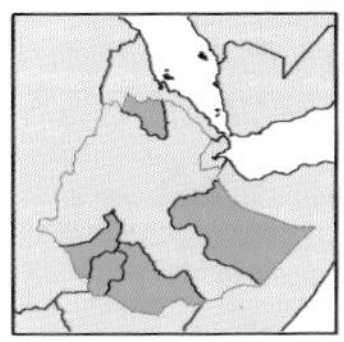

Habenaria cultriformis

The specific epithet '*cultriformis*' refers to the leaves which are shaped like plough-shares or knife blades. Fritz Kraezlin described it in 1893 from plants collected by Georg Wilhelm Schimper in Mettgalo and Amba Sea, Tigray.

Plant 20–68 cm tall. Stem leafy throughout length, leaves largest at base. Leaves to 18 × 2.3–4 cm, lanceolate, uppermost leaves bract-like. Inflorescence up to 18 × 4–5 cm, densely few- to many-flowered. Bracts 1.6–2.4 cm long, mostly longer than the pedicel with ovary, elliptic-lanceolate, slightly glandular. Flowers sweetly scented, greenish with yellow-green sepals. Dorsal sepal 4 × 1.8 mm, elliptic, concave, lying over the column; lateral sepals 9 × 6.5 mm, elliptic, with an obliquely set blunt apex, reflexed. Petals bipartite at the base; upper (posterior) lobe 3–5 × 0.5 mm, linear-curved, connate with dorsal sepal forming a hood, papillate; lower (anterior) lobe 14 × 4.2 mm at base, ovate below, tapering rather abruptly just above base, reflexed, decurved. Lip 3-lobed just above the base; side lobes 8 × 0.5 mm, linear, acute; midlobe 14 × 0.8 mm, linear; spur 1.6–2 cm long, dilated towards apex.

Habitat and distribution In grassland or open bushland between 1140 and 2200 m in Tigray, Harerge, Sidamo, Gamo Gofa and Kefa. Also in Yemen.

Flowering period May and June in Sidamo; August to October elsewhere in Ethiopia.

Conservation status Rare, possibly endangered.

Notes Differs from *H. tweediae* by the spur being 17–23 mm long rather than 28 mm or more long.

32. H. rivae *Kraenzl.*

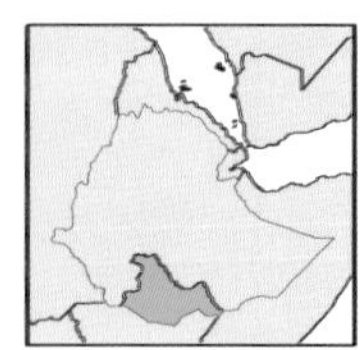

Habenaria rivae

The specific epithet '*rivae*' was given in honour of Italian collector Dominico Riva who collected the type specimen. Fritz Kraenzlin described it in 1897 from a plant collected near Giaribuli, Sidamo.

Plant to 120 cm tall. Stem leafy. Leaves 6, the largest oblong-lanceolate, up to 30 × 5 cm. Inflorescence erect, densely many-flowered. Bracts lanceolate, overtopping flowers, 2 cm long. Flower colour unknown; pedicel and ovary 12 mm long. Dorsal sepal cucullate, oblong, 6 mm long. Lateral sepals subobliquely ovate, 6 mm long. Petals ovate, 6 mm long, with a small angular anterior lobe, posterior lobe antheriferous. Lip simple, ovate-lanceolate, 6 mm long, with a central fleshy mid-line; spur filiform, inflated a little at apex, 18 mm long.

Habitat and distribution In Sidamo. Habitat and altitude unknown. Unknown elsewhere.

Flowering period August to October.

Conservation status Critically endangered.

Notes The flowers of the type are abnormal, the petals bearing anthers and the lip lacking side lobes. Further material is needed to ascertain the status of this taxon. *H. rivae* differs from the rest of the species in the section by the lip being entire rather than 3-lobed and the bracts overtopping flowers.

33. H. tweedieae *Summerh.*

Habenaria tweedieae

The specific epithet '*tweedieae*' is given in honour of the collector of the type specimen, Mrs Marjorie Tweedie, who collected the plant on Mt Elgon, Kenya. Victor Summerhayes described it in 1933.

Plant 40–100 cm tall. Stems erect, leafy. Leaves 9–15, the middle 2 to 5 lanceolate, elliptic-lanceolate or oblong-elliptic, the largest 8–25 × 2–5 cm, the upper ones bract-like. Inflorescence 10–45 × 4.5–6.5 cm, densely 15- to many-flowered. Bracts lanceolate, 1.5–3.8 cm long. Flowers green and white, scented; pedicel and ovary almost straight, 2.5–3.5 cm long, sparsely hairy. Dorsal sepal erect, convex, elliptic, 6–7 × 2.5–4 mm, somewhat glandular on outside. Lateral sepals reflexed, obliquely semi-orbicular, 8–10 × 4.5–6.5 mm. Petals bipartite; posterior lobe adnate to

Habenaria tweedieae

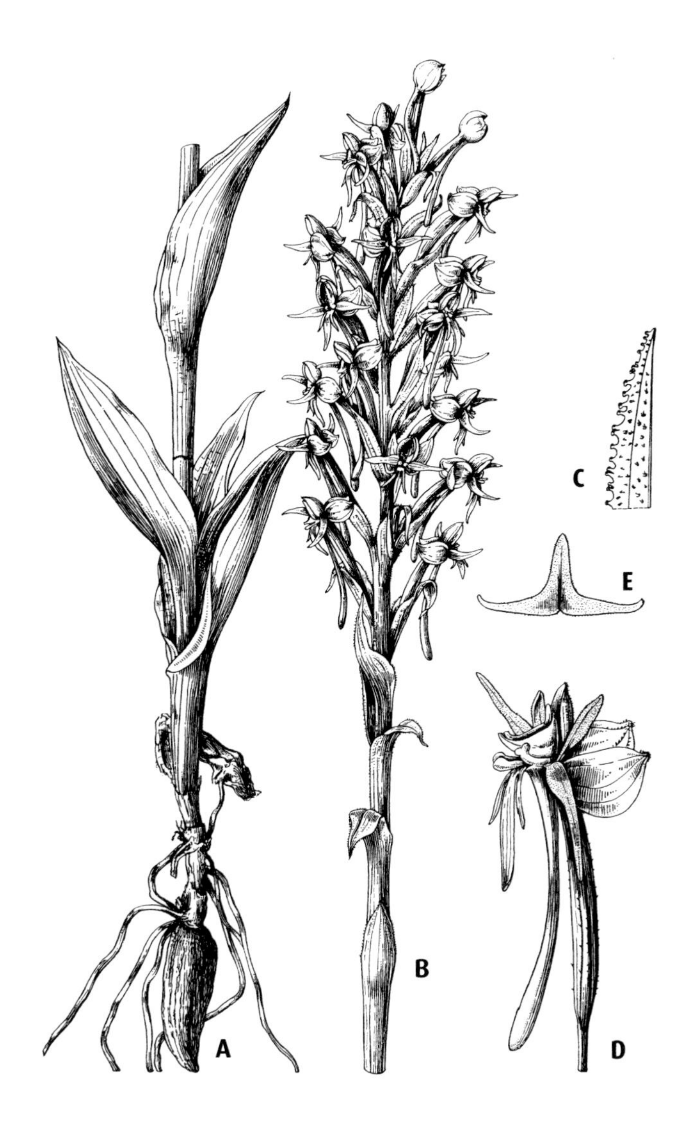

A **Tuber and stem,** × ½; **B** **inflorescence,** × ½; **C** **bract margin,** × 4; **D** **flower,** × 1½; **E** **rostellum,** × 2. All drawn by Stella Ross-Craig from *Tweedie* 25.

dorsal sepal, linear, puberulous, ciliolate, 6–7.5 × 1 mm; anterior lobe lanceolate-falcate, 9–11 × 1.5–3 mm, puberulous in lower part. Lip deflexed, 3-lobed to base; side lobes linear, 5–8.5 × 0.5 mm; midlobe linear, 10.5–12.5 × 1 mm; spur untwisted, slightly apically dilated, 25–36 mm long.

Habitat and distribution On roadsides in degraded juniper forest between 1950 and 2250 m in Sidamo, Gamo Gofa, Kefa and Wellega. Also in Uganda, Kenya and Tanzania.

Flowering period August and September; November.

Conservation status Local, but vulnerable.

Notes Differs from *H. cultriformis* by the spur being usually 28 mm or more long.

Section **Mirandae** *Summerh.*

A section distinguished by it large habit, leafy stems, bifid petals, erect dorsal sepal, hairy base to the lip and anthers over 6 mm long.

The section is represented by one species in Ethiopia.

34. H. rautaneniana *Kraenzl.*

The specific epithet '*rautaneniana*' was given in honour of Martti Rautanen who collected the type specimen from Namibia. Fritz Kraenzlin described it in 1902.

H. rautaneniana

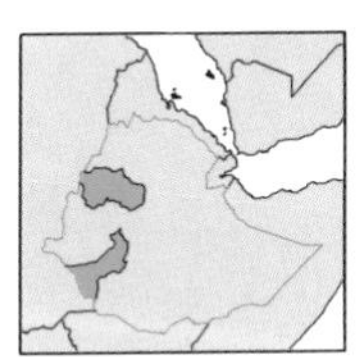

Habenaria rautaneniana

Plant 30–80 cm tall. Stem leafy throughout its length. Leaves 5–11, linear, the largest 18–33 × 7–16 mm, the upper 1 or 3 much shorter, similar to the bracts. Inflorescence 8–24 × 5.5–7 cm, loosely to rather densely 6–35-flowered. Bracts lanceolate, 1.2–3.3 cm long. Flowers green or yellow-green, with a disagreeable odour at night; pedicel and ovary 2.5–4 cm long. Dorsal sepal very convex, narrowly elliptical, 10–15 × 3–5.5 mm. Lateral sepals spreading, obliquely ovate or ovate-lanceolate, 11–18 × 6–8.5 mm. Petals bipartite nearly to the base; posterior (upper) lobe erect, adherent to the dorsal sepal, linear, 9–15 × 1 mm; anterior lobe curved upwards, linear, 14–23 × 0.5–1.5 mm. Lip projecting forwards or more or less deflexed, tripartite from an undivided base 3 mm long; midlobe linear, 12–17 × 1 mm; side lobes much diverging, ligulate or lanceolate-ligulate, 11–15 × 1.5–3 mm broad; all lobes, and especially the side lobes, pubescent in the basal half; spur recurved and much swollen in the basal half, 14–23 mm.

Habitat and distribution In boggy grassland in Gojam and Kefa at about 1800 m. Also in Tanzania, Angola, Malawi, Zambia, Zimbabwe and South Africa.

Flowering period August.

Conservation status Rare and possibly threatened in Ethiopia. Very local elsewhere.

Section **Ceratopetalae** *Kraenzl.*

A section in which most species are distiguished by the horn-like nature of their bifid petals and tri-lobed lip and by their 4–20 mm long anther canals. Four species have been reported from Ethiopia.

Key

1	Spur less than 40 mm long; stigmatic arms less than 12 mm long	2
–	Spur over 55 mm long; stigmatic arms over 13 mm long	3
2	Stigmatic arms 3–8 mm; side lobes of lip widest above base, often with teeth on outer margin	**35. H. cornuta**
–	Stigmatic arms 8–11 mm long; side lobes of lip narrowed from the base, entire	**36. H. clavata**
3	Spur 55–75 mm long	**37. H. holubii**
–	Spur 130–220 mm long	**38. H. cirrhata**

35. H. cornuta *Lindl.*

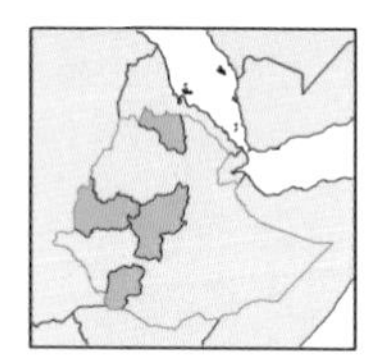

Habenaria cornuta

The specific epithet '*cornuta*' refers to the anterior lobes of the petals curling upwards like long horns. John Lindley described it from a plant collected by Jean Francois Drège in the Cape Province, South Africa. The species was also known by the name *H. ceratopetala* described by Achille Richard in 1840 from a plant collected by Richard Quartin-Dillon in Adwa, Tigray.

Plant 20–80 cm tall. Stem erect, leafy throughout its length. Leaves 9–15, linear-lanceolate to ovate, the largest 2–10 × 0.7–4.5 cm, decreasing in size upwards, similar to the bracts. Inflorescence 5–19 × 3–6 cm, loosely to densely 4- to many-flowered. Bracts lanceolate, 2–7 cm long, mostly shorter than the pedicel and ovary. Flowers pale green or yellow-green; pedicel and ovary 1.8–2.8 cm long. Dorsal sepal very convex and almost boat-shaped, 5–16 × 4–8 mm. Lateral sepals obliquely semi-orbicular, more or less rolled up lengthways, 6–16 × 5.5–10.5 mm. Petals bipartite nearly to the base; posterior (upper) lobe more or less erect, usually adherent to the dorsal sepal, linear, 6.5–14.5 × 0.5 mm; anterior lobe curling upwards like a long horn, linear below, subulate above, 2–4.5 × 1–2 mm. Lip deflexed, tripartite from an undivided basal part more or less 1 mm long; midlobe more or less incurved in the apical part, linear, 9.5–19 × 1 mm; side lobes somewhat diverging, narrowly lanceolate or lanceolate-linear from a broad base, often with 1 or 2 short teeth or shortly pectinate on the outer margin at about the middle, 8–18 × 1–2.5 mm towards the base; spur much swollen in the apical half or third, 14–27 mm long.

H. cornuta

Habitat and distribution In open, often badly drained, grassland or in *Brachystegia* woodland between 850 and 2400 m in Tigray, Shewa, Gamo Gofa and Wellega. Also in Nigeria to DR Congo, Uganda, Kenya, Tanzania, Malawi, Zambia, Zimbabwe and South Africa.

Flowering period August and September.

Conservation status Locally common.

Notes Differs from *H. clavata* by the side lobes of the lip often having teeth on the outer margin.

36. H. clavata (*Lindl.*) *Rchb.f.*

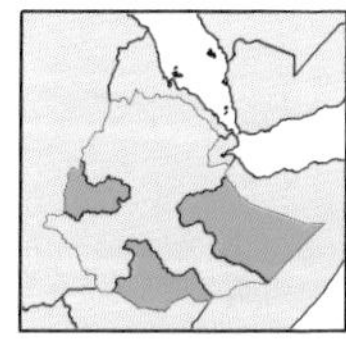

Habenaria clavata

The specific epithet '*clavata*' refers to the swollen, club-like spur. John Lindley described it as *Bonatea clavata* in 1837 from a South African collection made by Jean Francois Drège. The name was changed to *H. clavata* by H.G. Reichenbach in 1865.

Plant 20–80 cm tall. Stem leafy throughout its length. Leaves 8–13, ovate, elliptical-lanceolate or lanceolate, the largest 7–13 × 1.5–4 cm, the upper leaves smaller, similar to the bracts. Inflorescence 4–17 × 6–8.5 cm, rather lax, 5–16-flowered. Bracts lanceolate, 1.5–3.5 cm long, distinctly shorter than the pedicel with ovary. Flowers green with whitish centre; pedicel with ovary incurved, 3.5–4.5 cm long. Dorsal sepal very convex, ovate-elliptical, 11–19 × 5–8 mm. Lateral sepals deflexed, rolled up lengthways, obliquely obovate, 12–18 × 6–9 mm. Petals bipartite nearly to the base; posterior (upper) lobe erect, adherent to the

H. clavata

dorsal sepal, narrowly linear to filiform, 9–15 mm long; anterior lobe curved upwards like a long horn, linear, 25–40 × 1 mm. Lip projecting forwards, tripartite from an undivided base 2 mm long; midlobe often more or less incurved, linear, 17–23 mm long; side lobes linear, 12–18 mm long; all lobes scarcely 1 mm broad; spur more or less parallel to the ovary and pedicel, much swollen in the apical third or quarter, 25–40 mm long.

Habitat and distribution In *Acacia* woodland and upland grassland between 1100 and 2285 m in Wellega, Sidamo and Harerge. Also in Nigeria, Cameroon, DR Congo, Tanzania, Malawi, Zambia, Zimbabwe and South Africa.

Flowering period May in Sidamo; September elsewhere in Ethiopia.

Conservation status Vulnerable.

Notes Differs from *H. cornuta* by the side lobes of the lip being entire rather than toothed.

37. H. holubii *Rolfe*

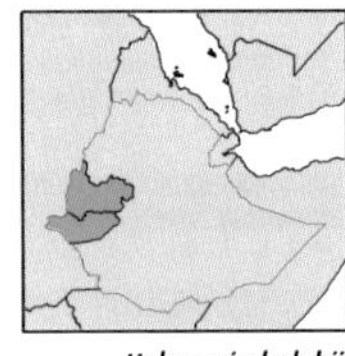

Habenaria holubii

The specific epithet '*holubii*' was given in honour of the collector of the type specimen, Emil Holub, who collected it from Zimbabwe. Robert Rolfe described it in 1898.

Plant 25–80 cm tall. Stem leafy throughout its length. Leaves 7–11, ovate to narrowly lanceolate, the largest 7–15 × 2–4.5 cm, upper ones smaller, similar to the lower bracts. Inflorescence 4–20 × 7–13 cm, loosely 3–19-flowered. Bracts lanceolate, 2–6 cm long. Flowers pale green or greenish-white; pedicel and ovary 4.5–7.5 cm long. Dorsal sepal very convex, 15–20 × 6–10 mm. Lateral sepals deflexed, rolled up lengthways, obliquely oblanceolate or semi-orbicular, 18–24 × 10–15 mm. Petals bipartite nearly to the base; posterior lobe erect, adherent to the dorsal sepal, narrowly linear, 15–20 × 0.5 mm; anterior lobe curving upwards and outwards like a horn, linear, tapering in the upper part, 20–40 × 2–3 mm. Lip tripartite from a basal undivided part 2 mm long; midlobe deflexed and incurved, linear, 20–35 × 1 mm; side lobes projecting forwards but upright on each side of the column, lanceolate or rarely lanceolate-linear, 6–10 × 1.5–3 mm; spur parallel to the ovary and pedicel, clavately swollen at the apex, 55–75 mm long.

Habitat and distribution In swampy grassland and in black sandy soils in grassland between 1000 and 1200 m in Illubabor and Wellega. Also in Guinea to Central African Republic, DR Congo, Uganda, Kenya, Tanzania, Angola, Zambia, Zimbabwe and Namibia.

Flowering period July and August.

Conservation status Endangered in Ethiopia but locally common elsewhere.

Notes Differs from *H. cirrhata* by the spur being less than 75 mm long.

38. H. cirrhata (*Lindl.*) *Rchb.f.*

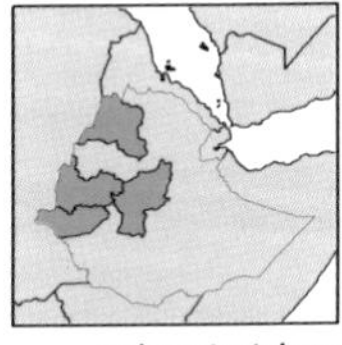
Habenaria cirrhata

The specific epithet '*cirrhata*' refers to the narrow, curled lobes of the petals and lip. John Lindley described it as *Bonatea cirrhata* in 1835 from a plant collected from Madagascar by Charles Lyell. H.G. Reichenbach transferred it to *Habenaria* in 1865.

Plant 50–130 cm tall. Stem leafy throughout its length. Leaves 9–16, almost orbicular to lanceolate, the largest 7–22 × 3.5–9 cm, the upper ones smaller, similar to the bracts. Inflorescence 4–28 × 11–15 cm, loosely 3–11-flowered. Bracts lanceolate, 2–6.5 cm long, usually distinctly shorter than the pedicel and ovary. Flowers green with white central parts; pedicel and ovary straight or somewhat curved, 5.5–8 cm long. Dorsal sepal elliptical-lanceolate, 2–2.5 × 7–10 mm. Lateral sepals deflexed, rolled up lengthwise, very obliquely obovate or semi-orbicular, 2–3 × 1–1.5 cm. Petals bipartite nearly to the base; posterior (upper) lobe more or less adherent to the dorsal sepal, narrowly linear, 2–2.5 × 0.5 mm; anterior lobe curved forwards and upwards like a horn, linear below, fleshy, 5–9 × 1 mm. Lip projecting forwards, tripartite from an undivided basal part 2–4 mm long; midlobe linear, 3–4 × 1 mm; side lobes more tapering and narrower, 2–3 cm long; spur swollen in the apical third or quarter, 13–22 × 3–4 mm.

Habitat and distribution In grassland with scattered bushes and *Brachystegia* woodland between 300 and 1600 m in Gonder, Shewa, Illubabor and Wellega. Also in Guinea Republic to Nigeria, Cameroon, DR Congo, Sudan, Uganda, Kenya, Tanzania, Malawi, Zambia and Madagascar.

Flowering period July.

Conservation status Very local in Ethiopia but locally common elsewhere.

Notes Differs from *H. holubii* by the spur being 130–220 mm long.

Section **Macrurae** *Kraenzl.*

A section in which the species are distinguished by their leafy stems, broadly lanceolate petal and lip lobes, long spur (over 10 cm), and 8–12 cm long stigma lobes. It is represented in Ethiopia by a single species.

39. H. perbella *Rchb.f.*

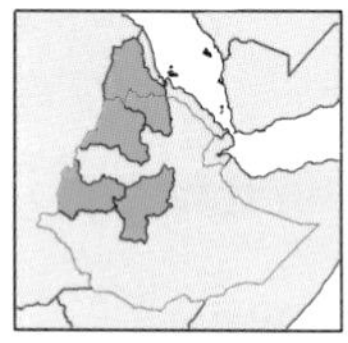

Habenaria perbella

The specific epithet '*perbella*' refers to the very beautiful flowers. H.G. Reichenbach described it in 1881 from a plant collected by Georg Wilhelm Schimper in Semien.

Plant to 60 cm tall. Leaves 7, all along the stem, up to 10 × 2.6 cm. Inflorescence 9 cm long, 3–8-flowered. Bracts similar to leaves. Dorsal sepal ovate, 14 × 8.5 mm. Lateral sepals obliquely ovate-oblong, 15.5–17 × 5–6 mm. Petals bipartite almost to the base; posterior lobe ovate-oblong, 10.5 × 3.5 mm; anterior lobe narrowly triangular, 14 × 2.6 mm, adnate to column for basal 2 mm. Lip 3-lobed with an undivided base 5 mm long; midlobe linear, 8.5–17 × 2.2–3 mm; side lobes linear-lanceolate, acute, 13.5 × 2.8 mm; spur pendent, 115–160 mm long.

Habitat and distribution Habitat unknown. The species is found between 1200 and 1500 m in Tigray, Gonder, Shewa and Wellega and in Eritrea. Unknown elsewhere.

Flowering period August.

Conservation status Vulnerable.

Notes May be conspecific with the widespread *H. walleri*.

Section **Diphyllae** *Kraenzl.*

A section in which the species are characterised by their 1 or 2, basal, orbicular or heart-shaped leaves appressed to the ground.
Seven species have been reported from Ethiopia.

Key

1	Leaves 2	2
–	Leaf solitary	5
2	Petals entire	**40. H. vaginata**
–	Petals bipartite	3
3	Lip lobes broad, obovate, obtuse or subacute	**41. H. macrura**
–	Lip lobes narrow, linear	4
4	Spur with a knee-like bend 3 mm from base; anterior petal lobe linear, 20 mm long, emerging well above base of posterior lobe	**43. H. decumbens**
–	Spur lacking angular bends; anterior petal lobe filiform, 30–40 mm long, emerging from base of posterior lobe	**42. H. armatissima**
5	Stem and sepals glabrous	**44. H. busseana**
–	Stem and sepals pubescent	6
6	Spur 7–10 mm long; dorsal sepal 3–4 mm long	**45. H. holothrix**
–	Spur 15 mm long; dorsal sepal 7 mm long	**46. H. keayi**

40. H. vaginata *A.Rich.*

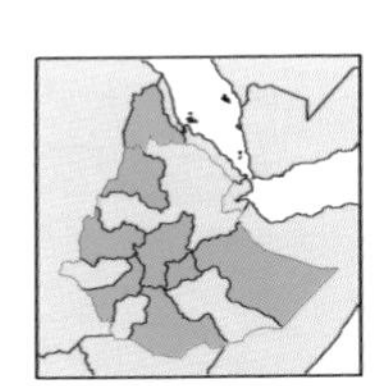
Habenaria vaginata

The specific epithet '*vaginata*' refers to the sheaths along the inflorescence axis. Described by Achille Richard in 1840 from a plant collected between Adwa in Tigray and Mensa in Eritrea by Richard Quartin-Dillon.

Plant 10–50 cm tall, with 2 leaves at or near the base and a number of sheaths along the rest of the stem. Leaves more or less appressed to the ground, the upper one sometimes 1–2 cm above the lower, the lower one reniform, orbicular or ovate, the upper one where present normally narrower, broadly ovate to lanceolate- ovate, 7–10 × 5–9 cm. Inflorescence 4–20 × 2–3 cm, rather densely 6- to many-flowered. Bracts lanceolate, 7–25 mm long, the lower ones longer than the ovary with pedicel. Flowers green; pedicel with ovary 1–1.5 cm long. Dorsal sepal very convex, elliptical, 3.5–5.2 × 2–5 mm. Lateral sepals obliquely and curved semi-ovate, 5–6.5 × 2–3 mm. Petals entire, erect, adherent to the dorsal sepal, obliquely oblong-lanceolate or lanceolate, 3.5–5.3 × 1.5–1.7 mm. Lip deflexed, tripartite from an undivided base 1 mm long; lobes more or less incurved, linear, the midlobe 4.5–7 × 1 mm, the side lobes a little shorter and narrower; spur 1.7–35 cm long, slightly swollen in the apical half.

H. vaginata

Habitat and distribution In short grassland, especially where damp, or at edge of forest, sometimes with bushes between 1500 and 2800 m in Gonder, Wellega, Shewa, Arsi, Harerge, Sidamo and Kefa and in Eritrea. Also in Kenya and Tanzania.

Flowering period April and May in Sidamo; July to October elsewhere in Ethiopia.

Conservation status Vulnerable.

Notes Differs from *H. macrura* by its smaller flowers and the petals being entire rather than bipartite. Note that some of the Ethiopian material shows considerable variation in the length of the spur, the size of the flowers, and the length of the rostellum midlobe.

41. H. macrura *Kraenzl.*

The specific epithet '*macrura*' refers to the flowers with a long spur. Fritz Kraenzlin described it in 1892 from a plant collected in Angola.

Plant 20–65 cm tall. Stem with 2 leaves at the base and a number of appressed sheaths almost covering its whole length. Basal leaves appressed to the ground, ovate, elliptic or nearly orbicular, rounded, only slightly cordate at the base, 2.5–9.5 × 2–5 cm. Inflorescence up to 16 × 4–6 cm, rather densely up to 11-flowered. Bracts lanceolate, 2–4 cm long, shorter than the pedicel and ovary. Flowers white or cream with sepals green outside, sometimes fragrant; pedicel and ovary 2.5–4.5 cm long. Dorsal sepal erect, ovate or broadly ovate, convex, 8–13 × 4.5–11 mm. Lateral sepals obliquely semi-ovate,

H. macrura

9–16.5 × 3.5–6 mm; all sepals with numerous cross-veins. Petals bipartite nearly to the base; posterior (upper) lobe erect, curved-lanceolate to obliquely semi-ovate, 7.5–11–5 × 2.5–7.5 mm; anterior lobe obliquely oblanceolate, 9.5–19 × 2.5–6 mm. Lip projecting forwards, tripartite from an undivided base 2–6.5 mm long; midlobe lanceolate-ligulate, 10.5–19 × 3–6.5 mm; side lobes 8.5–23 × 3–8 mm. Spur narrowly cylindrical, often more or less covered by the stem-sheaths, 9–17 cm long.

Habitat and distribution In grassland or *Combretum* woodland between 1200 and 2400 m. In Nigeria to DR Congo, Uganda, Tanzania, Angola, Malawi and Zambia. The species is said to occur in Ethiopia by Summerhayes (1968), but no Ethiopian specimens seen so far.

Flowering period Not known in Ethiopia. Flowers in August in the neighbouring Sudan.

Conservation status A widespread species that can be locally common elsewhere.

Notes Differs from *H. vaginata* by its larger flowers with bipartite petals.

42. H. armatissima *Rchb.f.*

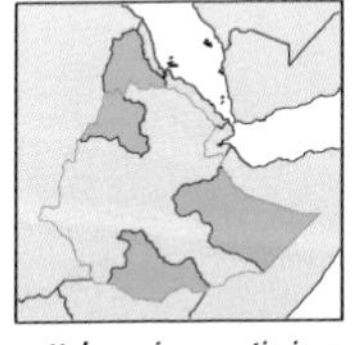
Habenaria armatissima

The specific epithet '*armatissima*' refers to the flowers, which appear to be armed with bristles. H.G. Reichenbach described it in 1881 from a plant collected in the Semien, northern Ethiopia.

Plant 30–70 cm tall. Stem with 2 large leaves at the base and several much smaller ones along its length.

Basal leaves opposite, appressed to the ground, broadly ovate to reniform, often somewhat cordate at the base, 6–18 × 7–22 cm. Inflorescence 10–20 × 8–12 cm, closely 10- to many-flowered. Bracts lanceolate, 1–3.5 cm long, distinctly shorter than the pedicel with ovary. Flowers white; pedicel with ovary 3.5–5 cm long. Dorsal sepal erect, elliptical-ovate, very convex, 11.5–16 × 5–8 mm. Lateral sepals reflexed, very obliquely semi-orbicular or semi-ovate, 13–17.5 × 4–7.5 mm. Petals bipartite nearly to the base; posterior (upper) lobe erect, curved-linear, 11.5–14.5 mm long, scarcely 1 mm broad; anterior lobe projecting forwards or spreading, almost filiform, 3–4 cm long, about as broad as the posterior in the lower part. Lip projecting forwards, tripartite from an undivided base more or less 2 mm long; all lobes linear or almost filiform, 0.5–1 mm broad, the midlobe 1.5–2 cm long, the side lobes 3–4.5 cm long; spur pendent, narrowly cylindrical, hardly swollen in the apical part, 8.5–21.5 cm long.

H. armatissima

Habitat and distribution In forest, open marshy ground, grassland by streams, deciduous thicket or mixed dry woodland between 300 and 1650 m in Gonder, Harerge and Sidamo and in Eritrea. Also in Mali, Cameroon, Sudan, Kenya, Tanzania, Mozambique, Zambia, Zimbabwe, Malawi and Namibia.

Flowering period May in Sidamo; August elsewhere in Ethiopia.

Conservation status Rare and vulnerable in Ethiopia. Locally common elsewhere.

Notes Differs from *H. decumbens* in having a straight rather than bent spur.

43. H. decumbens *S.Thomas & P.J.Cribb*

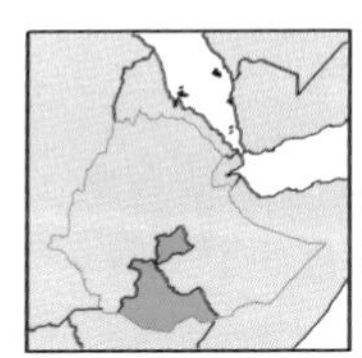
Habenaria decumbens

The specific epithet '*decumbens*' refers to the distinctive bend in the spur. Sarah Thomas and Phillip Cribb described it in 1995.

Plant 16–35 cm tall. Stem with 3–4 sheaths along its length, and two leaves at the base. Leaves adpressed to ground, broadly ovate, 5–7 × 4–5 cm. Inflorescence 8–9 × 4–5 cm, 5–8-flowered. Bracts ovate-oblong, up to 22 × 9 mm. Flowers pale green; pedicel with ovary 20–25 mm. Dorsal sepal ovate-oblong, convex, 17–18 × 11 mm. Lateral sepals obliquely ovate-lanceolate, 20 × 5.5 mm. Petals bipartite, basal undivided part 6 mm long; posterior lobe partly covered by dorsal sepal, narrowly triangular, 12.2 × 4.5 mm; anterior lobe narrowly linear, rounded, 20 × 1.2 mm. Lip 3-lobed, basal 3–4 mm triangular, undivided and fused to column; midlobe linear to narrowly triangular, 15 × 1.7 mm; side lobes exceeding midlobe, linear, 25 × 1.1 mm; spur 20–22 mm long, distinctive knee-like bend 3 mm above base, gently curved above this, slightly widened at the apex.

Habitat and distribution In grassland between 1900 and 2600 m in Arsi and Sidamo. Unknown elsewhere.

Flowering period July to November.

Conservation status Rare and critically endangered.

Notes Differs from *H. armatissima* by having a bent, rather than straight, spur.

44. H. busseana *Kraenzl.*

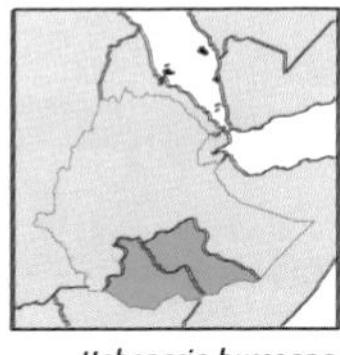
Habenaria busseana

The specific epithet '*busseana*' was given in honour of Walter Busse, who collected the type from Tanzania. Fritz Kraenzlin described it in 1902.

Plant 20–40 cm tall. Stem erect, glabrous, somewhat flexuous with a single leaf at the base. Leaf appressed to the ground, broadly ovate or reniform, 3–4 × 5–7 cm. Inflorescence 6–12 × 4 cm long, loosely 13–25-flowered. Bracts lanceolate, scarcely as long as the pedicel. Flowers green; pedicel and ovary 2.5–3 cm long. Dorsal sepal oblong-lanceolate, convex, 4–4.5 × 2.4–4 mm. Lateral sepals 5–7 × 3–4.5 mm. Petals bipartite nearly to the base; posterior lobe erect, lanceolate, hidden beneath the dorsal sepal and adherant to it at the base, 3–4 mm long; anterior lobe

Habenaria decumbens

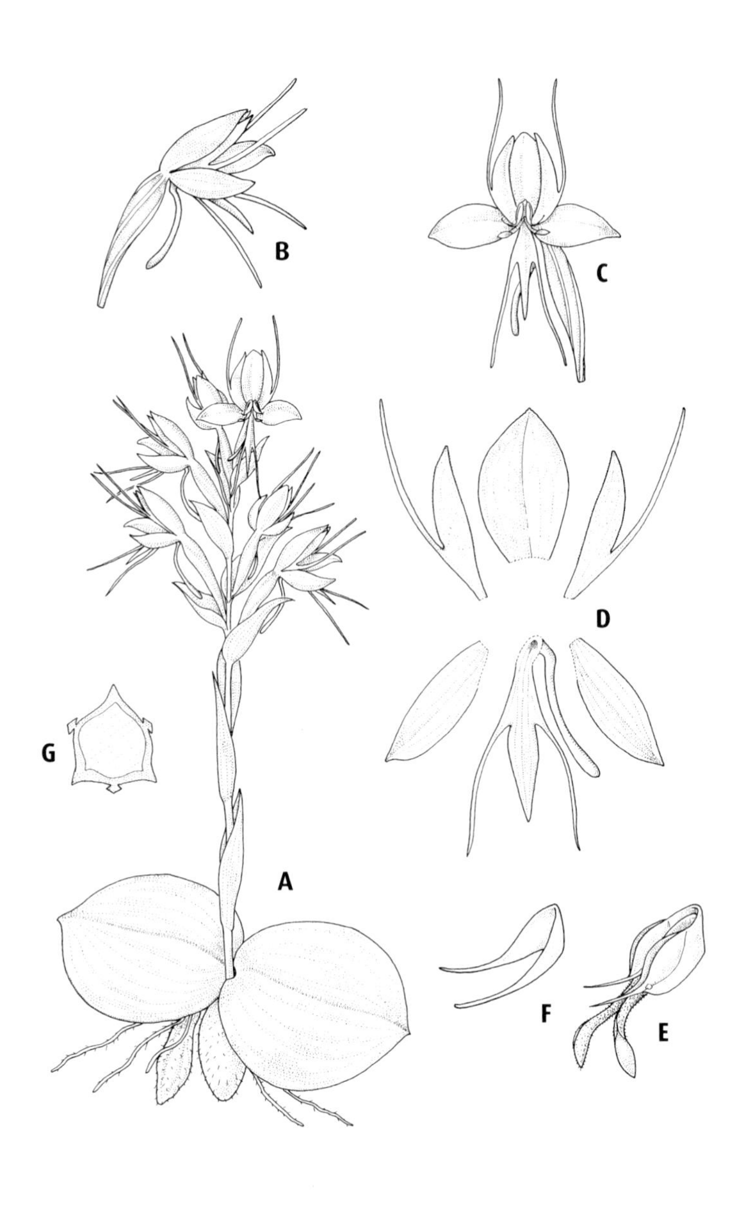

A habit × ½; **B** flower, front view × ⅔; **D** sepals and lip flattened × ½;
E column, side view × 2; **F** rostellum × 2; **G** ovary, cross section × 2.
All drawn from *de Wilde* 5582 by Susanna Stuart-Smith.

curved upwards, linear-lanceolate, 10 mm long. Lip tripartite nearly to the base; undivided part 1.5 mm long, lobes lanceolate, equal, 7–8 mm long; spur incurved, narrow at the base, swollen in apical part, 20–25 mm long.

Habitat and distribution In grassland between 1580 and 2450 m in Bale and Sidamo. Also in Tanzania.

Flowering period May and June.

Conservation status Rare and endangered in Ethiopia and probably throughout its range.

Notes Ethiopian specimens match the Tanzanian material well, but the disjunct distribution suggests that further information is needed about the Ethiopian material before its identity can be confirmed. It differs from *H. holothrix* and *H. keayi* by the stem and sepals being glabrous rather than pubescent.

45. H. holothrix *Schltr.*

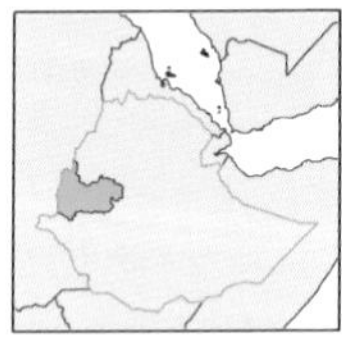

Habenaria holothrix

The specific epithet '*holothrix*' refers to the similarity to the orchid genus *Holothrix*. Rudolf Schlechter described it in 1903 from a plant collected in Angola by Hugo Baum.

Plant 10–30 cm tall. Stem with a single leaf at the base. Leaf adpressed to the ground, broadly ovate to almost orbicular, 1–3 × 1.5–3 cm, cordate at the base, glabrous or with a few scattered hairs on the upper surface, margins ciliate; sheaths up to 1 cm long, hairy. Inflorescence 1–11 × 1.5 cm, loosely 2–21-flowered. Bracts lanceolate, 3–7 mm long, softly hairy. Flowers green or yellow-green, sometimes fragrant; pedicel and ovary softly hairy, 7–11 mm long. Dorsal sepal convex, 3–4 × 2–3, spreading, hairy. Lateral sepals obliquely elliptical-lanceolate, hairy outside, 3.5–4.5 × 1.5 mm. Petals erect, 2-lobed or bipartite from below the middle, glabrous; posterior lobe curved-lanceolate, 2.5–3 × 1 mm; anterior lobe very variable in length, 0.5–5 × 0.25 mm. Lip curved downwards, tripartite from an undivided base 0.5–1 mm long, glabrous; all lobes linear, the midlobe 3–6 × 0.75 mm; the side lobes 1.5–5.5 × 0.25 mm; spur only slightly thickened in the apical half, 7–10 mm long.

Habitat and distribution In open *Combretum* woodland between 950 and 2000 m in Wellega. Also in Tanzania, Angola and Zimbabwe.

Flowering period December to February.

Conservation status Rare and possibly endangered.

Notes Differs from *H. keayi* by the shorter spur.

46. H. keayi *Summerh.*

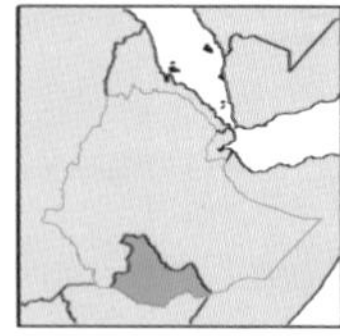

Habenaria keayi

The specific epithet '*keayi*' was given in honour of Ronald Keay who collected the type specimen in Nigeria. Victor Summerhayes described it in 1951.

Plant up to 25 cm tall. Stem densely pubescent, with single leaf adpressed to the ground. Leaf up to 7.5 × 10 cm, heart-shaped, upper side densely pubescent, lower side glabrous. Inflorescence 5–8 cm long, 8–17-flowered. Bracts lanceolate, 10 mm long, pubescent. Flowers green and white; pedicel and ovary pubescent, 13–15 mm long. Dorsal sepal erect, ovate, pubescent, 7 × 3 mm. Lateral sepals pubescent, 7.5 × 3 mm. Petals bipartite, glabrous, posterior lobe adnate to dorsal sepal, 7 × 1.5 mm; anterior lobe linear, curved upwards, 12 × 0.6 mm. Lip trilobed, glabrous; midlobe linear, reflexed, 10 × 1 mm; side lobes spreading, linear, 12 × 1 mm; spur swollen near apex, 15 mm long.

Habitat and distribution In *Combretum-Terminalia* wooded grassland in Sidamo. Also in Nigeria and Oman.

Flowering period June and July.

Conservation status Rare and possibly endangered. Known only from one collection in Ethiopia.

Notes Differs from *H. holothrix* by the longer spur.

Section **Trachypetalae** *Summerh.*

A section distinguished by its striking very hairy entire or partly divided petals. The lip, rostellum, anther and stigmata are also distinct.
A single species is found in Ethiopia.

47. H. longirostris *Summerh.*

The name comes from the long side lobes of the rostellum in the flower. Victor Summerhayes described it in 1932 based on a collection from Nigeria.

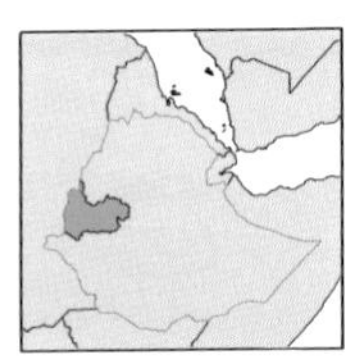
Habenaria longirostris

Plant up to 1 metre tall. Stem erect, stout, leafy. Leaves 6–13, broadly ovate to lanceolate-ovate, 7–16 cm long, 2–8 cm wide. Inflorescence 7–35 cm long, 6–8 cm in diameter, rather laxly many-flowered; bracts lanceolate, 1.5–4 cm long, shorter than or overtopping flowers. Flowers large, green with lower part white; pedicel and ovary 1.5–2.5 cm long. Dorsal sepal erect, lanceolate, 12–20 mm long, 3–4 mm wide at base, 2 mm in upper half. Lateral sepals spreading, obliquely lanceolate, 16–22 mm long, 4–5 mm wide. Petals bilobed in upper three-quarters, adnate to the dorsal sepal, ciliate; upper lobe attached to dorsal sepal, 13–20 mm long; lower lobe 18 mm long, ciliate. Lip 21–30 mm long, trilobed from an undivided 4–5 mm long base; mid lobe linear, 17–25 mm long, 1 mm wide; side lobes linear, 15–21 mm long, 1 mm wide; spur attached to ovary, 15–21 mm long.

H. longirostris

Habitat and distribution Bushland and bushy meadows in Wellega. Also reported from Uganda, Nigeria and Guinea Bissau.

Flowering period August.

Conservation status Endangered in Ethiopia. Rare and possibly endangered throughout its range.

Notes Its long-ciliate petals are very distinctive.

6. BONATEA *Willd.*

Terrestrial herbs with elongated fleshy and tuberous roots. Stems unbranched, usually very leafy. Leaves arranged all along the stem, but sometimes withered by the time the flowers are open. Inflorescence terminal, 1–many-flowered. Flowers resupinate, green or yellow and white. Dorsal sepal free, but usually forming a helm with the upper petal-lobes, the laterals partly united to the base of the lip, the anterior lobes and the stigmatic arms. Petals 2-lobed, the upper lobe usually adherent to the dorsal sepal, the lower (anterior) lobe adnate at the base to the stigmatic arm and the lip. Lip adnate in the basal part to the stigmatic arms and lateral sepals, the free part 3-lobed, spurred at the base; disk usually with a distinct tooth in front of the spur-opening; spur long or short, cylindrical. Anther upright, the loculi adjacent and parallel, canals usually more or less elongated, adnate to the side lobes of the rostellum, auricles undivided, rugulose; pollinaria 2, each with sectile pollinium, rather long slender caudicle and small naked viscidium; stigmatic processes elongated, the lower part adnate to the lip, the free part club-shaped; rostellum standing out in front of the anther, convex and usually hooded, 3-lobed, with a relatively short middle lobe and often long slender side lobes.

A genus of nearly 20 species, almost restricted to the mainland of Africa with one species in Arabia (Yemen). Two species are reported from Ethiopia.

Key

1 Spur 5–6.5 cm long; lip with claw 1–1.5 cm long, midlobe 15–20 mm long **2. B. rabaiensis**

\- Spur 10–21 cm long; lip with claw 1.5–3 cm long; midlobe 20–35 mm long **1. B. steudneri**

1. B. steudneri *(Rchb.f.) Th.Dur. & Schinz*

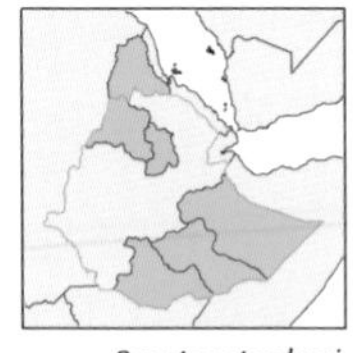
Bonatea steudneri

The specific epithet '*steudneri*' was given in honour of the collector of the type specimen, Dr Steudner, who collected the plant from Keren, Eritrea. H.G. Reichenbach described it as *Habenaria steudneri* in 1881. Théophile Durand and Hans Schinz transferred it to *Bonatea* in 1895.

Plant 25–125 cm tall, glabrous. Stem leafy throughout its length. Leaves 10–20, lanceolate, oblong-lanceolate or ovate-elliptical, the largest 7–19 ×

B. steudneri

3–5 cm, decreasing in size above the middle of the stem, the uppermost smaller, similar to the bracts. Inflorescence cylindrical or almost corymbose, 3–28 × 10–20 cm, rather loosely 3–30-flowered. Bracts 2–5 cm long, shorter than the pedicel and ovary. Flowers white and green; pedicel and ovary slender, 4–7 cm long. Dorsal sepal elliptical-lanceolate, 20–30 × 10–20 mm; laterals very obliquely broadly lanceolate. Petals bipartite almost to the base. Lip tripartite from a long narrow claw, 15–30 mm long; midlobe decurved and usually sharply bent back in the middle, linear, 2–3.5 × 1–3 mm; side lobes spreading, much longer and narrower than midlobe, 2.5–8.5 cm long, 1–2 mm broad; spur narrowly cylindrical with the apical part somewhat swollen, 10–21 cm long.

Habitat and distribution In bushland and scrub, at edges of thickets, rocky places in rainforest, rocky slopes between 1000 and 2800 m in Welo, Harerge and Sidamo and in Eritrea. Also in Arabia, Sudan, Somalia, DR Congo, Uganda, Kenya, Tanzania, Zambia and Zimbabwe.

Flowering period July in Sidamo; September to November elsewhere in Ethiopia.

Conservation status Data deficient. A widespread but rare species in Ethiopia.

Notes *Bonatea steudneri* differs from *B. rabaiensis* by the spur being 10–21 cm long.

2. B. rabaiensis *(Rendle) Rolfe*

The specific epithet '*rabaiensis*' refers to the Rabai Hills, near Mombasa, Kenya from where the type was collected. Alfred Rendle described it as *Habenaria rabaiensis* in 1895 from a plant collected by W.E. Taylor. Robert Rolfe transferred it to *Bonatea* in 1898.

Plant 2.5–5 cm tall, almost glabrous. Stem leafy along its entire length. Leaves 8–9, lanceolate, elliptical-lanceolate or oblanceolate, the largest 6–11 × 3–4.5 cm the uppermost much smaller, similar to the bracts. Inflorescence 5–11 × 6–11 cm, rather loosely 3–7-flowered. Bracts lanceolate, 1–2.5 cm long, usually less than half as long as the pedicel and ovary. Flowers green and white; pedicel and ovary straight, 3.5–5 cm long. Dorsal sepal ovate, 17–20 × 13 mm; laterals united to the lip and stigmatic arms for more or less 10 mm, very obliquely lanceolate-ovate, 20 × 10 mm. Petals bipartite nearly to the base; posterior (upper) lobe erect, adherent to the dorsal sepal, linear, 15–20 mm long, 1.5 mm broad; anterior lobe adnate to the lip for 7 mm, above this curved upwards and outwards, linear, narrowed upwards, 25–30 mm long. Lip deflexed, tripartite from a narrow basal part 10–15 mm long; midlobe linear-ligulate, 15–20 × 1–2 mm; side lobes linear, 25–30 × 1 mm; spur swollen in the apical half, 5–6.5 cm long.

Habitat and distribution No record from Ethiopia. In Kenya.

Flowering period Unknown for Ethiopia. May to June elsewhere.

Conservation status Data deficient.

Notes Said to occur in Ethiopia by Tournay (1972) but no Ethiopian specimens seen during the preparation of this account. It differs from *B. steudneri* by the spur being 5–6.5 cm long rather than 10–21 cm long.

B. rabaiensis

7. PLATYCORYNE *Rchb.f.*

Terrestrial herbs with fleshy or tuberous roots. Stems more or less leafy. Leaves scattered along the stem or mostly in a tuft at the base, usually narrow. Inflorescence terminal, 1–many-flowered. Flowers resupinate, usually orange or yellow or greenish, rarely white. Sepals free, the laterals deflexed, the dorsal forming a helm with the 2 petals. Petals usually adherent to the dorsal sepal, usually simple but rarely with a short lobe at the base of the front margin. Lip free, spurred, simple or with short side lobes; spur cylindrical. Column erect, the anther upright, loculi parallel, contiguous, canals varying in length, rarely scarcely developed, adnate to the side lobes of the rostellum; pollinaria 2, each with sectile pollinium, a long caudicle and elliptical viscidium; stigmatic processes distinct, decurved, thick with rounded knob-like apices, the rostellum either between the anther-loculi or standing out in front of the anther, 3-lobed, more or less erect, the middle lobe often overtopping the anther, the side lobes porrect and usually projecting beyond the hood, sometimes short and shoulder-like. Capsules oblong or fusiform.

A genus of 17 species, restricted to the mainland of Africa except for one species in Madagascar. A single species is found in Ethiopia.

Platycoryne crocea (*Rchb.f.*) *Rolfe*

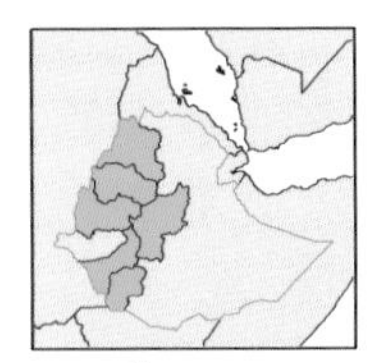
Platycoryne crocea

The specific epithet '*crocea*' refers to its yellow-orange flowers. H.G. Reichenbach described it as *Habenaria crocea* in 1878 from plants collected south of Lesi and near Matluoli, Sudan by Georg August Schweinfurth. Robert Rolfe transferred it to *Platycoryne* in 1898.

subsp. montiselgon (*Schltr.*) *Summerh.*

The epithet refers to Mt Elgon in Kenya where the type was collected.

Plant 10–40 cm tall. Stem with a tuft of leaves at the base and several more scattered along its length. Leaves 5–12(–16), linear or linear-lanceolate, the largest 15–80 × 2–8 mm, the upper ones at intervals on the stem, more or less appressed, lanceolate, acute, shorter than the basal ones. Inflorescence short, 0.5–5 × 2–3.5 cm, densely 2–9-flowered. Bracts lanceolate, 7–18 mm long, usually shorter than the pedicel and ovary. Flowers yellow or orange; pedicel and ovary 14–19 mm long. Dorsal sepal broadly ovate, 6.5–12 ×

P. crocea

P. crocea

3.5–7.5 mm. Lateral sepals obliquely oblong-lanceolate, 5.5–11.5 × 2–3.5 mm. Petals adherent to the dorsal sepal, obliquely ligulate-lanceolate, 4.5–10.5 × 2 mm. Lip 3-lobed at the base, altogether 5–12 mm long; midlobe ligulate, 1–2 mm broad, the side lobes almost tooth-like, up to 2 mm long; spur swollen in the apical part, 9–15 mm long.

Habitat and distribution Grassy areas in shallow soil over rocks or lava pavements between 1200 and 2350 m in Gonder, Gojam, Shewa, Gamo Gofa, Kefa and Wellega. Also in Sudan, Uganda and Kenya.

Flowering period June to August.

Conservation status Rare and vulnerable.

Notes *Platycoryne crocea* subsp. *crocea* from Kenya differs in having 3 basal leaves not much longer than cauline ones; dorsal sepal 6–7 mm long; side lobes of lip less than 0.5 mm long. Subsp. *ochrantha* from Tanzania differs in having 2–6 cauline leaves and triangular side lobes of the lip which are 0.3–1 mm long.

8. ROEPEROCHARIS *Rchb.f.*

Terrestrial herbs with tuberous roots. Stems leafy. Leaves usually narrow, scattered along the stem. Inflorescence terminal, few- to many-flowered. Flowers resupinate, green. Sepals free, the laterals spreading. Petals free, simple but often toothed or irregular, rather fleshy. Lip free, spurred, 3-lobed or rarely simple; spur cylindrical. Column erect, the 2 anther-loculi separate at the sides of the broad connective and more or less divergent, canals scarcely developed, auricles large, elliptical or quadrate; pollinaria 2, each with a sectile pollinium, a long caudicle and small viscidium; stigmatic processes distinct, each 2-lobed, one lobe projecting downwards in front of the lip-base, the other upright in front of the anther-connective, rostellum 3-lobed, the side lobes spreading, narrowed towards the apex, the middle lobe low and rounded or emarginate, adnate to the connective.

A genus of five species, restricted to eastern Africa. Three species have been reported from Ethiopia.

Key

1	Lip entire; spur pendent at first then sharply curved upwards in apical half, upwardly-curved part greatly inflated	**1. R. alcicornis**
–	Lip tripartite; spur pendent, cylindrical or subclavate, gently curved or straight	2
2	Spur bifid at apex; leaves not mucronate; petals very obliquely lanceolate and almost S-shaped, twisted and folded in the middle, acute, 5.5–11.5 × 2.5–5.5 mm	**2. R. bennettiana**
–	Spur not bifid at apex; leaves mucronate; petals obliquely falcate-ovate, acuminate, 5–6 × 2 mm	**3. R. urbaniana**

1. R. alcicornis *Kraenzl.*

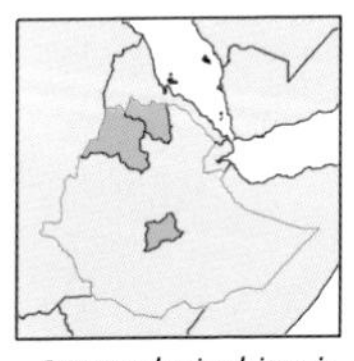
Roeperocharis alcicornis

The specific epithet '*alcicornis*' refers to the petals shaped like elk's horns. Fritz Kraenzlin described it in 1892 from a plant collected in Ethiopia by Georg Wilhelm Schimper.

Plant to 40 cm. tall. Stem leafy. Leaves 5, linear or narrowly lanceolate, the largest 11 × 1 cm, decreasing in size up the stem, the uppermost often similar to the bracts. Inflorescence narrow, 8 × 2 cm, rather densely 15-flowered. Bracts narrowly lanceolate, up to 15 × 5 mm, usually longer than the pedicel and ovary. Flowers greenish yellow; pedicel and ovary almost

R. bennettiana

straight, 10–13 mm long. Dorsal sepal ovate, 5.0 × 3.1 mm, 3-nerved. Lateral sepals very obliquely lanceolate-ovate, 6.5 × 3.8 mm. Petals obliquely lanceolate, margins wavy, crenulate, twisted and folded in the middle, 5.6 × 4.1 mm. Lip entire, very narrowly triangular, 10 × 2.5 mm; spur pendent at first then sharply curved upwards in apical half, upwardly curved part greatly inflated, 7.5 mm long.

Habitat and distribution In undulating county devoid of tree growth, small creeks with marshy borderlands on bottom of valley at about 2600 m in Gonder, Tigray and Arsi. Unknown elsewhere.

Flowering period September.

Conservation status Very local and endangered.

Notes *R. alcicornis* differs from the related species, *R. bennettiana* and *R. urbaniana* by the lip being entire rather than tripartite.

2. R. bennettiana *Rchb.f.*

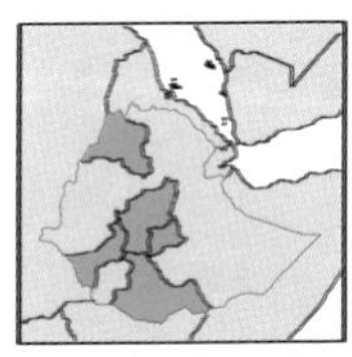

Roeperocharis bennettiana

The specific epithet '*bennettiana*' refers to John Bennett, Keeper of Botany at the British Museum (Natural History) in London. H.G. Reichenbach described it in 1881 from a plant Georg Wilhelm Schimper collected in Jan Meda, Gonder.

Plant 35–95 cm tall, almost entirely glabrous. Stem leafy throughout its length. Leaves 5–10, linear or narrowly lanceolate, the largest 11–25 × 1–2.5 cm, decreasing in size up the stem, the uppermost often similar to the bracts. Inflorescence narrow, 9–27 × 2–3 cm, densely or rather densely 15-many-flowered. Bracts narrowly lanceolate, 1–2.5 cm long, usually longer than the pedicel and ovary. Flowers green; pedicel and ovary almost straight, 10–13 mm long. Dorsal sepal erect,

R. bennettiana

broadly lanceolate or lanceolate-ovate, 6–11.5 × 4–6.5 mm, 5-nerved; laterals very obliquely lanceolate-ovate, 6.5–12.5 × 4–6.5 mm, 5–6-nerved. Petals very obliquely lanceolate and almost S-shaped, twisted and folded in the middle, 5.5–11.5 × 2.5–5.5 mm. Lip 3-lobed; undivided basal part oblong, 2.5–4.5 mm long; midlobe linear, 7–16.5 × 1–1.7 mm; side lobes divergent, linear, tapering towards the apex, 4.5–8.5 × 0.5–1 mm; spur cylindrical, shortly bifid at the apex, 5.5–14.5 mm long.

Habitat and distribution In swamps and damp grassland between 2000 and 3000 m in Gonder, Shewa, Arsi, Sidamo and Kefa. Also in Kenya, Tanzania, Malawi and Zambia.

Flowering period July to December.

Conservation status Vulnerable.

Notes *R. bennettiana* differs from *R. urbaniana* by the spur being bifid at apex.

3. R. urbaniana *Kraenzl.*

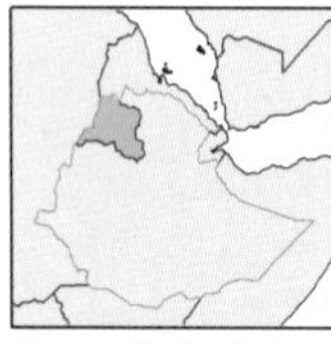

Roeperocharis urbaniana

The specific epithet '*urbaniana*' commemorates the Belgian botanist Ignatz Urban. Fritz Kaenzlin described it in 1892 from a plant collected in Mt Gunna, Gonder by Georg Wilhelm Schimper.

Terrestrial herb to 35 cm tall. Leaves 4–6, oblong-lanceolate or lanceolate, 5–12 × 10–15 mm. Inflorescence lax, up to 7 cm long, 8–13-flowered. Bracts elliptic to narrowly triangular, exceeding pedicel and ovary. Dorsal sepal ovate, 7 × 5 mm. Lateral sepals falcate-ovate, 9–10 × 3.5 mm. Petals entire, obliquely falcate-ovate, 5–6 × 2 mm. Lip 3-lobed in apical two thirds; basal undivided part 4 × 2.5 mm; midlobe linear, 9 × 1 mm; side lobes linear, 8 × 1 mm; spur subclavate, 15 mm long.

Habitat and distribution In grassy meadows with shrubs of *Hypericum* and *Erica* at about 2750 m in Gonder. Unknown elsewhere.

Flowering period August and September.

Conservation status Critically endangered.

Notes *R. urbaniana* differs from *R. bennettiana* by the spur being undivided at its apex.

9. DISA *Bergius*

Terrestrial herbs with tuberous roots. Stems unbranched, leafy. Leaves scattered along the flowering stem or on separate sterile shoots. Inflorescence terminal, 1- to many-flowered. Flowers resupinate, variously coloured. Sepals free, the dorsal erect, hooded or helmet-shaped, usually spurred, the laterals more or less spreading. Petals at the base more or less adnate to the column, often included in the dorsal sepal, variable in shape. Lip usually small and narrow, entire, not spurred. Column short; anther erect, horizontal or reflexed, the loculi parallel; pollinaria 2, each with sectile pollinium, caudicle and naked viscidium; stigmas united into a cushion below the rostellum, rostellum small, more or less 3-lobed, the middle lobe small, folded, the side lobes short, fleshy, often adnate to the petals. Ovary twisted, almost terete. Capsule cylindric, club-shaped or narrowly ellipsoid.

A genus of about 130 species, occurring mostly on the African mainland, predominantly South African, with four species in Madagascar and the Mascarene Islands. Represented by seven species in Ethiopia.

Key

1	Spur pendent	2
–	Spur erect or suberect	5
2	Flowers orange-scarlet, flesh-coloured or crimson	3
–	Flowers mauve, purple or pink	4
3	Flowers almost hidden by the bracts	**5. D. cryptantha**
–	Flowers much exceeding the bracts	**6. D. facula**
4	Petals bilobed in upper half; dorsal sepal erect, orbicular, obovate or ovate from a very short narrowed base, convex, 6–14 × 5–11 mm	**4. D. scutellifera**
–	Petals entire; dorsal sepal incurved, broadly elliptical, rounded, very convex, 4–6 × 3.5–5.5 mm	**2. D. deckenii**
5	Spur hooked at apex; flowers deep blackish purple	**7. D. hircicornis**
–	Spur not hooked at apex; flower colour not as above	6

D. facula

6 Petals broader than long, 3 × 4 mm, transversely oblong, truncate and obscurely bilobed on front margin **3. D. pulchella**

– Petals longer than broad, 4–6 × 2.5 mm, curved, oblanceolate or oblanceolate-oblong, acute, the front margin slightly widened at base, basal half attached to the column by a thin keel or ridge **1. D. aconitoides** subsp. **goetzeana**

---

1. D. aconitoides *Sond.*

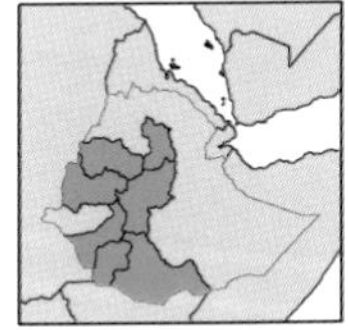

Disa aconitoides

The specific epithet '*aconitoides*' refers to the hooded flowers which resemble a small *Aconitum* (Ranunculaceae) flower. Described by Otto Sonder in 1847 from a plant collected by Christian Ecklon and Carl Zeyher in the Cape Province of South Africa.

subsp. **goetzeana** (*Kraenzl.*) *Linder*

The subspecific epithet commemorates the German collector Wilhelm Goetze. Fritz Kraenzlin described it in 1906 as *Disa goetzeana*.

Plant 20–65 cm tall. Sterile stems very short, with 2–3 overlapping sheaths and 1–2 leaves at the apex; leaves linear-oblanceolate, 90–120 × 7–10 mm. Flowering stems leafy throughout their length; leaves 4–8, lanceolate, the largest 4–10 × 7–30 mm. Inflorescence 7–28 × 2–4 cm, 10- to many-flowered, the rhachis spotted purple. Bracts narrowly lanceolate, 6–25 mm long, usually shorter than the flowers. Flowers whitish to purple, with purple or darker purple spots or markings; pedicel and ovary straight, 7–12 mm long, the ovary spotted with purple. Dorsal sepal 6–10 mm long, tapering towards the apex, base of sepal to apex of spur 6–10 mm. Lateral sepals obliquely oblong or elliptical. Petals inside the dorsal sepal, curved, oblanceolate or oblanceolate-oblong, the front margin slightly widened at base, 4–6 × 1–2 mm, basal half attached to the column by a thin keel or ridge. Lip projecting forwards, narrowly oblong or elliptical from a narrower base, 5–7 × 1.5 mm.

Habitat and distribution In grassland, sometimes with scattered shrubs, or *Brachystegia* woodland between 1310 and 2800 m in Gojam, Shewa, Sidamo, Gamo Gofa, Kefa and Wellega. Also in DR Congo, Uganda, Kenya, Tanzania, Malawi, Zambia and Zimbabwe.

Flowering period April to July.

Conservation status Rare in Ethiopia. Locally common elsewhere.

Disa hoods (dorsal sepals)

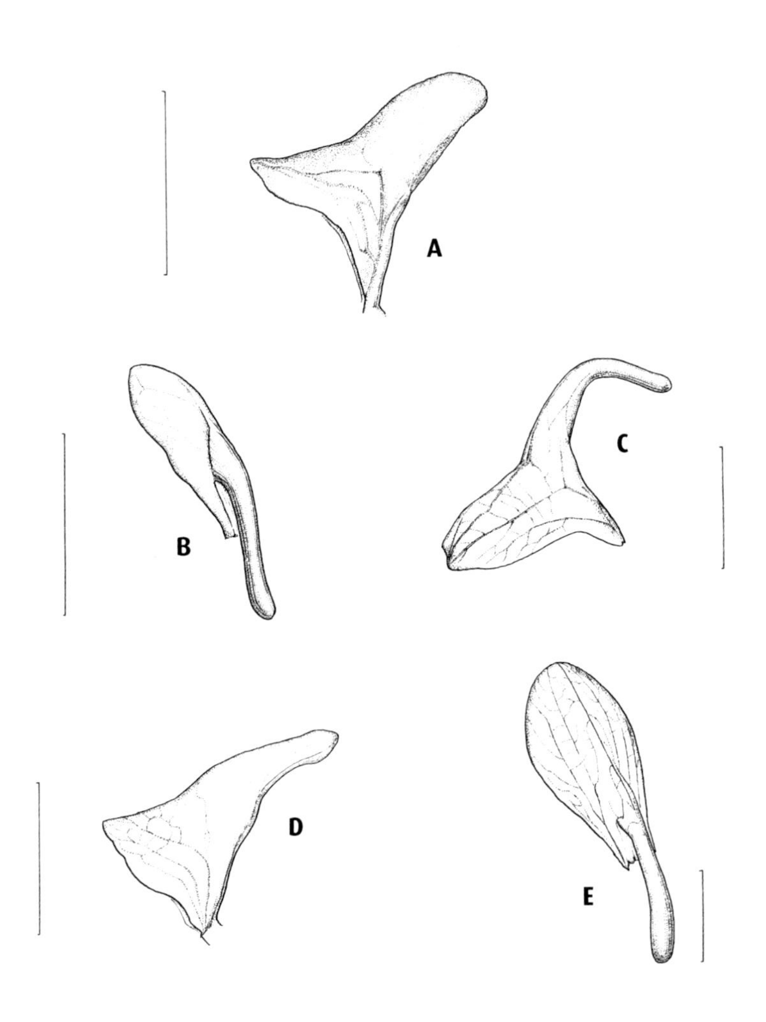

A *Disa aconitoides* subsp. *goetzeana*; **B** *D. cryptantha*; **C** *D. hircicornis*; **D** *D. pulchella*; **E** *D. scutellifera*. Scale bar = 5 mm.

Notes *D. aconitoides* differs from *D. pulchella* by the petals being longer than broad rather than broader than long.

2. D. deckenii *Rchb.f.*

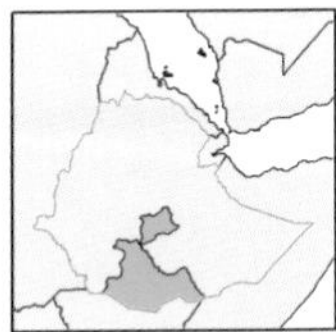

Disa deckenii

The specific epithet '*deckenii*' refers to Herr Decken, a colleague of Kersten on the expedition to Kilimanjaro. H.G. Reichenbach described it in 1881 from plants collected in Kilimanjaro, Tanzania by Kersten.

Plant 5–55 cm tall. Sterile stems to 4 cm long. Flowering stems with 7–13 lanceolate leaves, the largest 3–9 × 1–1.5 cm. Inflorescence cylindrical, 2–7 × 1.5–2.5 cm, 10- to many-flowered. Bracts lanceolate, 9–18 mm long. Flowers pink or rose-coloured, rarely crimson or mauvish-pink; pedicel with ovary almost straight, scarcely 1 cm long. Dorsal sepal broadly elliptical, spurred below the middle, 4–6 × 3.5–5.5 mm; spur cylindrical, 2.5–6 mm long. Lateral sepals spreading, obliquely elliptical or oblong-elliptical, 5.5–6.5 × 3–4 mm. Petals erect, lanceolate, 3.5–6 × 1.2–1.4 mm. Lip pendent, ligulate or ligulate-oblanceolate from a narrower base, 3–5 × 1 mm broad.

Habitat and distribution In glades at upper edge of forests, grassy spots in upland moor amongst *Erica* at about 2600 m in Arsi and Sidamo. Also in DR Congo and Sudan, Uganda, Kenya and Tanzania.

Flowering period June to October.

Conservation status Rare and vulnerable.

Notes *D. deckenii* differs from *D. scutellifera* by the petals being entire rather than bilobed in the upper half.

D. deckenii

3. D. pulchella *Hochst. ex A.Rich.*

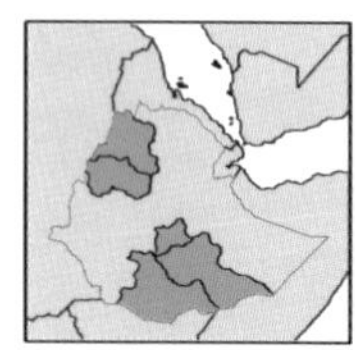
Disa pulchella

The specific epithet '*pulchella*' refers to the small and beautiful nature of the flowers. Achille Richard described it in 1851 from a plant collected in Enchet Kab, Semien, Gonder by Georg Wilhelm Schimper. The name was first used by the German botanist Christian Hochstetter.

Plant 12–35 cm tall. Stem leafy. Leaves 5–8, up to 13 × 2 cm, largest in centre of stem, lanceolate to linear-elliptic. Inflorescence cylindric, 3.5–11 × 2 cm, densely 3-many-flowered. Bracts about 1.6 cm long, lanceolate. Flowers sweetly scented, red-violet to purple with white petals. Dorsal sepal 5 × 5 mm, spurred in the middle on the dorsal surface; spur cylindric, 4–6 mm long. Lateral sepals 7 × 4 mm, oblong-elliptic. Petals 3–3.3 × 4–4.7 mm, hidden inside dorsal sepal, transversely oblong, truncate and obscurely bilobed on front margin. Lip 5.5 × 1.2 mm, linear or oblanceolate.

Habitat and distribution In *Erica* thickets and alpine grassland, scrub, riverbanks and woodland between 1800 and 3800 m in Gonder, Gojam, Arsi, Bale and Sidamo. Also in Yemen.

Flowering period May, July to September.

Conservation status Vulnerable.

Notes *D. pulchella* differs from *D. aconitoides* by the petals being broader than long rather than longer than broad.

4. D. scutellifera *A.Rich.*

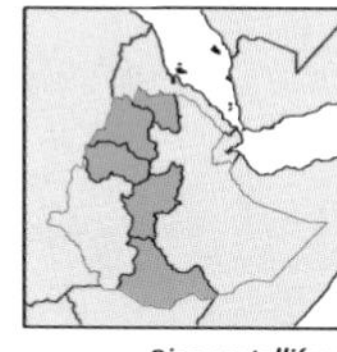
Disa scutellifera

The specific epithet '*scutellifera*' refers to the shield-shaped dorsal sepal. Achille Richard described it in 1840 from a plant collected by Richard Quartin-Dillon on Mt Selloda, near Adwa, Tigray.

Plant 25–75 cm tall. Sterile stems up to 13 cm long, covered by several overlapping sheaths, 1–2-leaved at the apex; leaves lanceolate or narrowly lanceolate, 10–30 × 1–2 cm. Flowering stems with 9–13 lanceolate leaves, the largest 7–12 × 1–2 cm. Inflorescence 5–19 × 2.5–4.5 cm, 14–many-flowered. Bracts lanceolate, 2–3.5 cm long, the lower ones often longer than the flowers. Flowers pink, bright carmine or mauve, often with darker spots, rarely white;

pedicel and ovary straight, 1.5 cm long. Dorsal sepal orbicular, obovate or ovate from a very short narrowed base, spurred below the middle, 6–14 × 5–11 mm; spur often slightly swollen in the apical part, 4.5–10 mm long. Lateral sepals obliquely oblong or elliptical, 8–13 × 4–6 mm. Petals erect, 2-lobed in the upper part, 6.5–11 mm long; lower (front) lobe obliquely elliptical with a somewhat cordate base, 5–7 × 3.5–5 mm; upper (back) lobe much smaller, linear-oblong, 3.5–5.5 × 1–1.5 mm. Lip pendent, linear, 6.5–10.5 × 0.7–1.7 mm.

Habitat and distribution In damp grassland or grassy rocky slopes between 1800 and 2850 m in Tigray, Gonder, Gojam, Shewa and Sidamo. Also in Sudan, Uganda and Kenya.

Flowering period May to July.

Conservation status Vulnerable.

Notes *D. scutellifera* differs from *D. deckenii* by the petals being bilobed in the upper half rather than entire.

D. scutellifera

5. D. cryptantha *Summerh.*

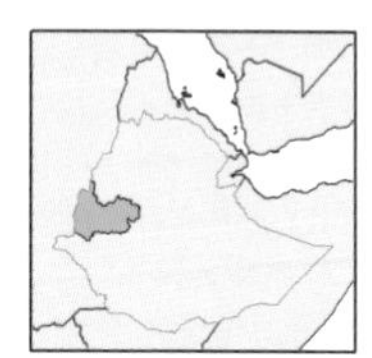
Disa cryptantha

The epithet '*cryptantha*' refers to the flowers which are almost hidden by the bracts. Victor Summerhayes described it in 1964 from a collection made by the Kew botanists Edgar Milne-Redhead and Peter Taylor in the Songea District of Tanzania.

Plant 30–50 cm tall, glabrous except for the roots. Sterile stems up to 4 cm long, with several red-spotted or tinged overlapping sheaths and a solitary apical leaf; leaf linear, acute, up to 160 × 5 mm wide. Fertile stems erect, slender. Leaves sheathing at base, appressed to the stem above with tips somewhat spreading, the largest 4–7 × 1 cm. Inflorescence 7–10 × 2 cm, rather densely 15–30-flowered; bracts leafy, almost covering the flowers, lanceolate, 1.8–3 cm long. Flowers flesh-coloured to crimson with dark purple spots; pedicel and ovary 1.2–1.3 cm long. Dorsal sepal erect, concave, broadly ovate to elliptic-ovate, 4–4.5 × 3–4 mm, spurred at the base; spur pendent, cylindrical, 3 mm long. Lateral sepals spreading, oblong to oblong-lanceolate, 5 × 3.5 mm. Petals erect, obovate, shortly bilobed at apex on inner margin, 5 × 3.5 mm. Lip pendent, linear, 4–5 × 1 mm.

Habitat and distribution In seasonally wet grassland between 950 and 1450 m in Wellega. Also in Zambia, Malawi, Zimbabwe and Tanzania.

Flowering period July to August.

Notes Distinguished by its small wine-red flowers in which the dorsal sepal has a slender cylindrical spur that is longer than the dorsal sepal, and petals which are bilobed, the posterior lobe being much smaller and shorter than the anterior one. The flowers are characteristically shorter than the bracts which have slender, elongate, pointed tips.

6. D. facula *P.J.Cribb, C.Herrm. & Sebsebe*

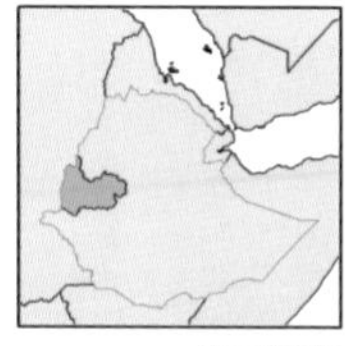
Disa facula

The specific epithet '*facula*' refers to the inflorescence which resembles a small burning torch from a distance. Described in 2003 from a specimen collected in Benishangul-Gumuz by Christof Herrmann.

Plant 60–75 cm tall. Sterile stems leafy, 2- to 3-leaved, drying blackish red, leaves suberect, lanceolate to oblanceolate, acute, up to 37 × 2.4 cm. Fertile stems erect, leafy, drying blackish reddish; leaves sheathing stem, bract-like, lanceolate, acuminate, 7–11 × 2.5–3 cm. Inflorescence densely many-flowered, 11–18 × 5.5–6 cm; bracts lanceolate, tapering rather abruptly in the middle and acuminate above, 3–3.6 cm long, almost as long as the flowers. Flowers showy, orange-red with red-spotted dorsal sepal; ovary more or less sessile, 1.8–2.2 cm long. Dorsal sepal erect, concave, elliptic, obtuse, 11–14.5 × 5–6 mm; spur pendent, cylindrical, slightly swollen at apex, 8–12 mm long. Lateral sepals spreading like wings, suberect, concave in apical part, obliquely oblong-ovate, acute, 16–18 × 5–6 mm. Petals bipartite, erect, anterior lobe concave, obliquely elliptic, obtuse, 7 × 6 mm; posterior lobe longer than the anterior lobe, enclosed within concave dorsal sepal, elliptic, obtuse, 3–4 × 2 mm. Lip pendent, linear, 11 mm long.

Habitat and distribution In seasonally wet grassland between 1400 and 1500 m in Wellega. Unknown elsewhere.

Flowering period July and August.

Conservation status Very rare and critically endangered.

Notes The flowers are larger and differently coloured from *D. scutellifera*, which grows at higher elevations in Ethiopia. The flowers are similar in colour to those of *D. roeperocharoides*, a species from south-central Africa, but they are larger.

D. facula

7. D. hircicornis *Rchb.f.*

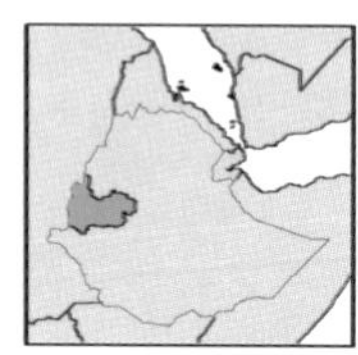

Disa hircicornis

The specific epithet '*hircicornis*' refers to the spur of the dorsal sepal, which is like a goat's horn. H.G. Reichenbach described it in 1881 from a collection made by John Kirk on Sochi Hill in Malawi.

Plant up to 85 cm tall, glabrous except for the roots. Sterile stems short, covered by overlapping red-spotted sheaths below, 1- to 3-leaved at apex. Leaves narrowly lanceolate or oblanceolate, 13–34 × 1–1.5 cm. Fertile stems erect, terete, leafy throughout length. Leaves 6–15, lanceolate or linear-lanceolate, 7–31 × 1–3.5 cm, the lowermost 1–3 sheathing and red-spotted or tinged. Inflorescence cylindrical, 6–25 × 2–3 cm; bracts lanceolate, 1–3 cm long. Flowers deep purple; pedicel and ovary 1 cm long. Dorsal sepal hooded, narrowly conical running into a recurved spur above, up to 13 × 5–8 mm wide. Lateral sepals spreading downwards, oblong-elliptic, 5–7 × 2.5–4 mm. Petals erect, inside the dorsal sepal, obliquely more or less rectangular, obtuse, 4–5.5 × 1.5–2.5 mm. Lip porrect or deflexed, linear, rounded, 4–5.5 × 1.5 mm.

D. hircicornis

Habitat and distribution In seasonally wet grassland by river at about 1490 m in Wellega. Also throughout tropical Africa from Nigeria to Angola and Zimbabwe.

Flowering period July to September.

Conservation status Very rare and vulnerable in Ethiopia but widespread and locally common elsewhere.

Notes Readily recognised by its cylindrical inflorescence of small, deep purple flowers with a dorsal sepal that has a crozier-like hook at its tip. Unlikely to be confused with any of the other Ethiopian species.

10. SATYRIUM *Sw.*

Terrestrial herbs with sessile undivided tubers. Stems unbranched, leafy, or covered with leaf-like sheaths. Foliage leaves either towards or at the base of the stem, or on separate sterile shoots. Inflorescence terminal, few to many-flowered. Bracts often reflexed. Flowers not resupinate, variously coloured. Sepals more or less united to the petals and the lip, the intermediate linear or ligulate, the laterals obliquely semi-elliptical or oblong. Petals more or less spathulate, linear or lanceolate, more or less united to the sepals at the base. Lip erect, more or less hooded, with a broad or narrow mouth, the apex sometimes recurved, usually with 2 spurs at base, sometimes with 2 extra shorter ones, rarely spurs absent; spurs long and slender to short and obtuse. Column erect and more or less incurved, included in the lip; anther pendent from the front of the column, the loculi parallel; pollinaria 2, each with a sectile pollinium, caudicle, and usually separate naked viscidium, rarely the 2 viscidia are united to form 1; stigma forming the upper lobe of the column, flat or hooded, rostellum projecting forward between the stigma and anther-loculi, more or less 3-lobed or 3-toothed, the middle lobe sometimes longer than side lobes. Ovary not twisted, often 6-ribbed. Capsules usually ellipsoid.

A genus of over 100 species, occurring mostly on the African mainland, predominantly South African, with five species in Madagascar and two in Asia. Represented by seven species in Ethiopia.

Key

1	Foliage leaves borne on flowering stem	2
–	Foliage leaves borne on separate, sterile stems; leaves on flowering stems sheath-like	5
2	Flowering stem with 1 or 2 basal heart-shaped or orbicular leaves appressed to the ground; flowers large, white, more than 2 cm in diameter	**1. S. aethiopicum**
–	Flowering stem with a number of leaves, those near the base not appressed to the ground; flowers small, green, red, brown or pink, less than 1 cm in diameter	3
3	Spurs shorter than the lip	**3. S. breve**
–	Spurs longer than the lip	4

S. crassicaule

4	Flowers pink; sepals and petals projecting forwards	**5. S. crassicaule**
–	Flowers green or yellow; sepals and petals sharply deflexed	**7. S. schimperi**
5	Flowers green or greenish-purple; spurs tapering to the apex	**2. S. brachypetalum**
–	Flowers red; spurs rounded at the apex	6
6	Spurs 11–17 mm long	**4. S. coriophoroides**
–	Spurs 8.5–10.5 mm long	**6. S. sacculatum**

1. S. aethiopicum *Summerh.*

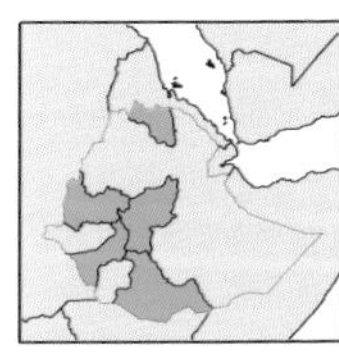
Satyrium aethiopicum

The specific epithet '*aethiopicum*' refers to Ethiopia, where the type originated. Victor Summerhayes described it in 1958 from a plant collected by Richard Quartin-Dillon on Mt Selloda, near Adwa, Tigray.

Plant up to 60 cm tall. Stem with 2 leaves. Leaves basal, adpressed to the ground, very broadly-ovate to orbicular or heart-shaped, up to 10 × 10 cm. Inflorescence 6–13 cm long, broadly cylindrical, rather densely 3–15-flowered. Bracts 12–34 × 6–12 mm, obovate. Flowers white; ovary 10–15 mm long. Dorsal sepal 12–16 × 3–4 mm, narrowly obovate. Petals similar to the dorsal sepal but slightly shorter and narrower. Lateral sepals 14–16 × 5–7 mm, oblong-oblanceolate. Lip excluding the spurs 14–16 × 14–16 mm, cucullate, broadly ovate to orbicular, obscurely 3-lobed; midlobe small, sub-triangular; side lobes rounded, enclosing column; spurs dorsal, 14–20 mm long, tapering to the apex.

Habitat and distribution On open hill sides on steep slopes and with low shrubs between 1600 and 2500 m in Tigray, Shewa, Sidamo, Kefa and Wellega. Unknown elsewhere.

Flowering period June to early October.

Conservation status Endangered.

Notes *S. aethiopicum* differs from the other Ethiopian species by the flowering stems having 1 or 2 basal heart-shaped or orbicular leaves adpressed to the ground rather than many leaves along the stem.

2. S. brachypetalum *A.Rich.*

The specific epithet '*brachypetalum*' refers to the short petals. Achille Richard described it in 1850, based on a collection from Ethiopia by Richard Quartin-Dillon.

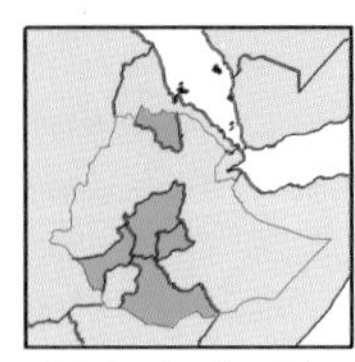

Satyrium brachypetalum

Plant up to 35 cm tall. Sterile stem up to 7 cm long, 2–5-leaved, the upper 1–2 up to 15 × 2.6 cm, narrowly oblong to oblanceolate or lanceolate. Fertile stem slender to robust, almost covered with more or less sheath-like leaves; leaves 4–19 × 1.6–3 cm, elliptic to lanceolate above. Inflorescence 4.5–15 cm long, narrowly cylindrical, densely many-flowered. Bracts 12–24 × 3–6 mm, lanceolate. Flowers suberect, dull reddish green to brown or green tinged with red-brown; ovary 7–9 mm long, papillose. Dorsal sepal 2.2 × 0.7 mm, similar to the petals, oblong. Lateral sepals 2 × 0.5 mm, oblong-oblanceolate. Petals 3 × 1.8 mm broad, obovate. Lip 6 × 4 mm, cucullate, reflexed at apex; spurs dorsal, 11–14 mm long, tapering to the apex.

Habitat and distribution In grasslands, amongst scrubby vegetation, on slopes between 2000 and 3200 m in Tigray, Welo, Shewa, Arsi and Sidamo. Also in Yemen.

Flowering period July to September.

Conservation status Endangered.

Notes *S. brachypetalum* differs from *S. coriophoroides* and *S. sacculatum* by the flowers being green or greenish purple rather than red.

3. S. breve *Rolfe*

The specific epithet '*breve*' refers to the short spurs at the back of the flower. Robert Rolfe described it in 1898 from a plant collected in Shire Highlands, Malawi by Buchanan.

Plant 10–70 cm tall. Sterile stem up to 19 cm long, 4-leaved; upper leaves lanceolate, up to 31 × 2 cm. Flowering stem leafy along its entire length; leaves 5–8, lanceolate or narrowly lanceolate, up to 6–21 × 1–4 cm, the uppermost much smaller. Inflorescence pyramidal, 2–13 × 2–5 cm, densley 10–many-flowered. Bracts lanceolate, 7–35 mm long, hairy inside. Flowers pink, deep red, mauve or purple, sometimes with darker markings; ovary 6–8 mm long. Sepals and petals united with one another in their lower half or third and with the lip in their lower third or quarter. Dorsal sepal narrowly oblong-oblanceolate, 7.5–22 × 1.5–3 mm. Lateral sepals obliquely oblong-elliptical, 2.5–7 mm broad. Petals similar to dorsal sepal but narrower; all tepals pubescent inside at the base. Lip very convex with a wide mouth, 7–16 mm long; spurs short and broad, 1.5–5.5 × 2 mm.

Habitat and distribution No authenticated record for Ethiopia. In East and South-central Africa.

Flowering period November to February in Tanzania.

Conservation status Locally common elsewhere in its range.

Notes Said to occur in Ethiopia by Tournay (1972) but no Ethiopian specimens seen during the preparation of this account. *Satyrium breve* differs from *S. crassicaule* and *S. schimperi* by the spurs being much shorter than the lip.

S. breve

4. S. coriophoroides *A. Rich.*

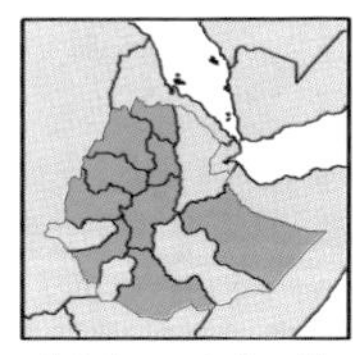

Satyrium coriophoroides

The specific epithet '*coriophoroides*' refers to the resemblance to the European orchid *Orchis coriophora*. Achille Richard described it in 1840 from a plant collected by Richard Quartin-Dillon on Mt Selloda near Adwa, Tigray.

Plant 40–100 cm tall, glabrous. Sterile stem to 5 cm high, 4–5-leaved; upper 1–2 leaves lanceolate to elliptical-lanceolate, up to 7–16 × 2.5–4.5 cm. Fertile stem slender to rather robust, almost covered by appressed sheathing leaves; sheaths lanceolate, acute, up to 4–10 × 1–3 cm. Inflorescence narrowly cylindrical, 8–23 × 1.5–2.5 cm, rather densely many-flowered. Bracts lanceolate, 1–2.5 cm long. Flowers white to crimson; ovary 5 mm long. Sepals and petals deflexed and more or less rolled up, united to one another and to the lip in the basal quarter. Dorsal sepal oblong-elliptical, 5–6.5 × 1 mm. Lateral sepals very obliquely elliptic-oblong, 5–6.5 × 1.5–2.5 mm. Petals oblong-oblanceolate, 4.5–6 × 0.5–1 mm, margins more or less ciliolate. Lip almost globular, with a narrow mouth, the apex shortly reflexed, 5–7 mm long; spurs parallel to the ovary, 11–17 × 1–2 mm.

Habitat and distribution In grassland between 1800 and 2850 m in Tigray, Gonder, Gojam, Welo, Shewa, Harerge, Sidamo, Kefa and Wellega. Also in Kenya and Cameroon.

Flowering period July to December.

Conservation status Vulnerable.

Notes *S. coriophoroides* differs from *S. sacculatum* by the spurs being 11–17 mm long rather than 8.5–10.5 mm long.

5. S. crassicaule *Rendle*

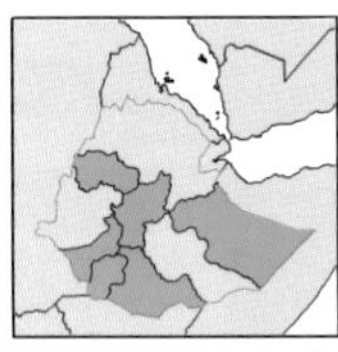
Satyrium crassicaule

The specific epithet '*crassicaule*' refers to the thick stem. Alfred Rendle described it in 1895 from a specimen collected by George Scott Elliot near the River Ruimi, Ruwenzori, Uganda.

Plant 30–120 cm tall, glabrous. Stem leafy along its entire length. Leaves 8–13, broadly lanceolate to ligulate or narrowly ligulate, up to 8–48 × 2–7.5 cm. Inflorescence cylindrical, 5–37 × 2–3.5 cm, densely many-flowered. Bracts lanceolate, 1–4 cm long, the lower ones often longer than the flowers. Flowers pink to mauve or rarely white; pedicel and ovary 1 cm long. Sepals and petals united to one another and to the lip in their basal third; intermediate sepal narrowly oblanceolate-elliptical, 4.5–8 × 1.5–2.5 mm. Lateral sepals obliquely oblong-elliptical, equalling the intermediate but a little broader. Petals similar to the intermediate sepal, 4–7 × 1.5 mm; all tepals 2–3-veined. Lip very convex and hooded with a shortly pointed apex and a rather broad mouth, 5–7.5 mm long and rather broader when spread out, 2-spurred at base; spurs slender, 8–13 × 1.5 mm.

Habitat and distribution In grassland with mixed scrub and scattered juniper between 1500 and 2700 m in Gojam, Shewa, Harerge, Sidamo, Gamo Gofa, Kefa and Wellega. Unknown elsewhere.

Flowering period April to June; September to December.

Conservation status Rare in Ethiopia but locally abundant elsewhere.

Notes *S. crassicaule* differs from *S. schimperi* by its taller habit and pink rather than green or yellow flowers.

6. S. sacculatum *(Rendle) Rolfe*

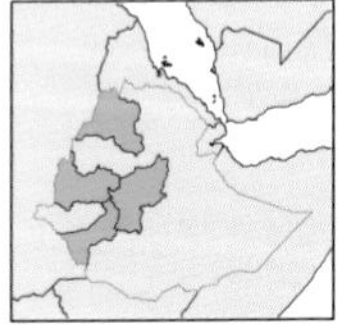

Satyrium sacculatum

The specific epithet '*sacculatum*' refers to the sac-like additional spurs. Alfred Rendle described it as *S. coriophoroides* A.Rich. var. *sacculatum* in 1895 based on a collection by George Scott Elliot from the Butahu valley, Ruwenzori and Nyamwamba Valley, DR Congo.

Plant 30–120 cm tall, glabrous. Sterile stem up to 7 cm long, 3–6-leaved; upper 1–2 leaves lanceolate to elliptic or elliptical-ovate, up to 6–24 × 1–7 cm. Fertile

S. sacculatum

stem slender to robust, almost entirely covered by more or less appressed sheathing leaves; sheaths 13–17, lanceolate, acute, up to 6–11 × 1.5–4 cm, the upper ones smaller and similar to the bracts. Inflorescence narrowly cylindrical, 7–38 × 1.5–2.5, densely many flowered. Bracts lanceolate, usually longer than the flowers. Flowers red to orange-brown, rarely white; ovary thick, 5 mm long. Sepals and petals united to one another and to the lip in their basal third. Dorsal sepal elliptical-oblong, 4.5–7 × 1 mm. Lateral sepals obliquely oblong-oblanceolate, a little longer than the intermediate, 1.5–2 mm broad. Petals oblong-oblanceolate, 4.5–6.5 × 1 mm, margins ciliolate, 1-veined. Lip almost spherical, the mouth very narrow, only 2 mm broad, the apex very shortly recurved, 5.5–7.5 mm long; spurs 8.5–10.5 × 1.5 mm, usually with a pair of additional very short spurs in front.

Habitat and distribution In wet meadows and in *Eucalyptus* plantations between 1600 and 2050 m in Gonder, Shewa, Kefa and Wellega. Also in Uganda, Kenya, Tanzania, Cameroon, DR Congo, Rwanda, Sudan, Zambia and Malawi.

Flowering period June to September.

Conservation status Endangered.

Notes *S. sacculatum* differs from *S. coriophoroides* by the spurs being 8.5–10.5 mm long rather than 11–17 mm long.

7. S. schimperi *A.Rich.*

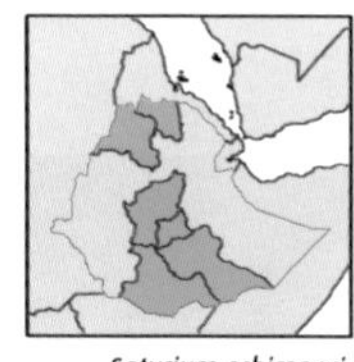
Satyrium schimperi

The specific epithet '*schimperi*' was given in honour of Georg Wilhelm Schimper, one of the collectors of the type. Achille Richard described it in 1851 from plants collected in Shire, Tigray by Richard Quartin-Dillon and Entchedtab, Semien, Gonder by Schimper.

Plant 15–60 cm tall, glabrous. Stem with two nearly opposite leaves towards the base. Leaves 5–7, the lowest 1–2 short, sheathing, the middle two more or less spreading, elliptical to elliptical-lanceolate, up to 4–16 × 2–5 cm, the upper 2–3 distant, lanceolate, the largest 30–80 × 8–14 mm. Inflorescence narrowly cylindrical, 5–22 × 1.5–2 cm, rather closely 9–to many-flowered. Bracts lanceolate, acute, 8–28 mm long. Flowers green or yellow-green; ovary 5–7 mm long. Sepals and petals with papillose margins, sharply decurved, united to one another and to the lip in their basal third. Dorsal sepal oblanceolate-oblong, 4–5 × 1 mm. Lateral sepals oblong-oblanceolate, twisted, 5 × 1.5 mm. Petals similar to lateral sepals but only 4 mm long. Lip fleshy, ellipsoid, hooded, 5–6 mm long; spurs parallel to ovary, 5–8 mm long, slender.

S. schimperi

Habitat and distribution In grassland amongst herbs and shrubs, in grassy places in open forest of juniper and *Erica arborea*, often in volcanic soils between 2000 and 3500 m in Tigray, Gonder, Shewa, Arsi, Bale and Sidamo. Also in Tanzania, Burundi, DR Congo, Angola, Malawi, Zambia and Zimbabwe.

Flowering period July to November.

Conservation status Vulnerable.

Notes *S. schimperi* differs from *S. crassicaule* by the flowers being green or yellow rather than pink.

11. DISPERIS *Sw.*

Erect mostly small terrestrial herbs arising from small tubers. Stems with 1–several sheathing scale-leaves at the base. Leaves 1–few, alternate or opposite, rarely almost obsolete. Flowers small, mostly under 2–5 cm long, white, yellow, green, pink or magenta, solitary or in several–many-flowered racemes; bracts leaf-like. Dorsal sepal united with the petals to form a structure which varies from an almost flat limb to an elongate spur; lateral sepals each with a conspicuous spur or

D. dicerochila

pouch near the inner margin (lacking in one South African species). Petals variously shaped, often falcate, obliquely acute or lobed at the apex, sometimes auriculate at the base. Lip remarkably modified, its claw joined to the face of the column and ascending above it, variously curving into the spur if present, often dilated into a smooth or papillate, straight or reflexed limb and usually bearing a simple or 2-lobed appendage which varies greatly in shape from species to species. Column erect, mostly stout; rostellum large, membranous, 2-lobed, produced in front into 2 rigid cartilaginous arms (fitting into the lateral sepal-pouches when in bud) holding the glands of the pollinia at their apices; anther-bearing part of column horizontal or ascending; anther-loculi distinct, parallel, more or less approximate; pollinia-granules secund in a double row on the margins of the flattened caudicles which curl up in a spiral on removal; staminodes present in some species; stigma 2-lobed, the lobes situated on either side of the adnate claw of the lip. Capsule cylindrical or ovoid, ribbed.

A genus of approximately 75 species extending from Togo and Ethiopia through tropical Africa to South Africa, the Mascarene Islands, India and thence to New Guinea. Represented by six species in Ethiopia.

Key

1	Leaf solitary, prostrate	**2. D. crassicaulis**
–	Leaves 2 or more, borne along stem	2
2	Leaves opposite	**3. D. dicerochila**
–	Leaves alternate	3
3	Hood longer than broad, not deeply saccate or conical; basal appendage of lip deeply bilobed and papillose, apical appendage papillose, circular and with a basal raised callus	**6. D. johnstonii**
–	Hood broader than long; lip not as above	4
4	Dorsal sepal recurved, obliquely bluntly conical	**4. D. galerita**
–	Dorsal sepal erect, conical	5
5	Dorsal sepal 5–7 mm long; lip with linear claw 4 mm long	**5. D. meirax**
–	Dorsal sepal 10–22 mm long; lip with linear claw 8–15 mm long	**1. D. anthoceros**

Disperis hoods and lips

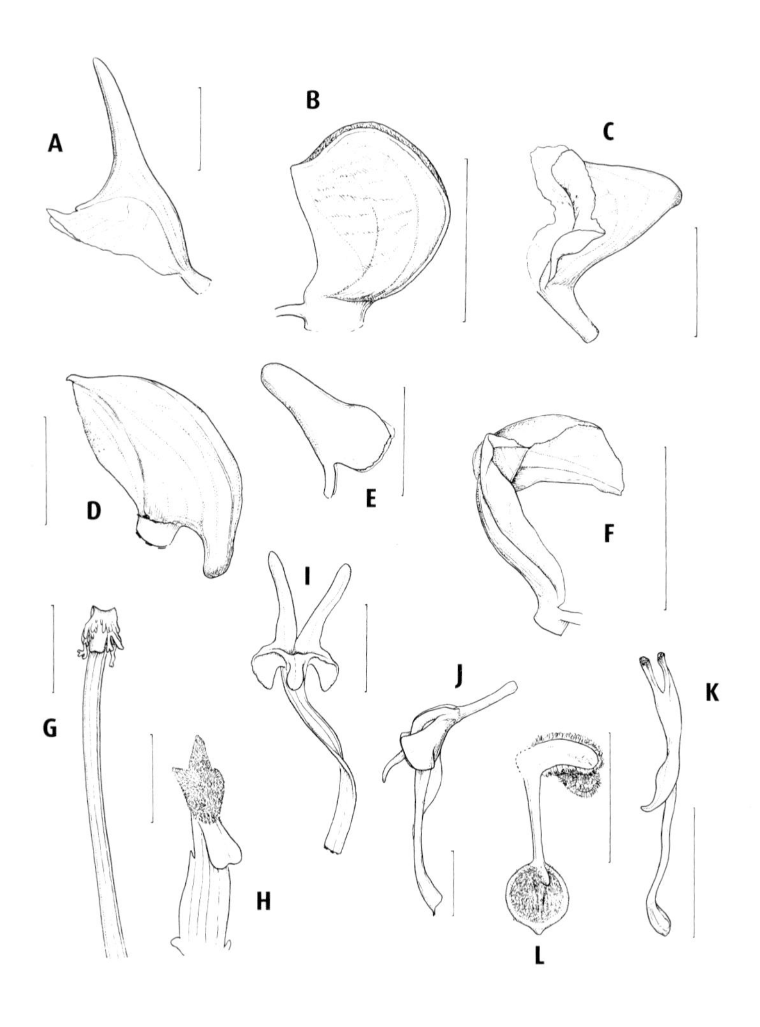

A *D. anthoceros;* **B** *D. crassicaulis;* **C** *D. dicerochila;* **D** *D. galerita;* **E** *D. meirax;* **F** *D. johnstonii;* **G** *D. anthoceros;* **H** *D. crassicaulis;* **I** *D. dicerochila;* **J** *D. galerita;* **K** *D. johnstonii;* **L** *D. meirax.*
A–E Scale bar = 5 mm. G–L Scale bar = 2 mm.

1. D. anthoceros *Rchb.f.*

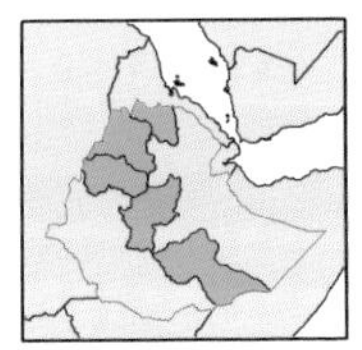
Disperis anthoceros

The specific epithet '*anthoceros*' comes from the Greek for a horned flower, in reference to the horn-like spur on the dorsal sepal. H.G. Reichenbach described it in 1881 from a plant collected in Tigray/Gonder by Georg Wilhelm Schimper.

var. **anthoceros**

Plant 8–30 cm tall. Leaves 2, opposite, ovate, 1.3–3.8 × 1–4 cm. Inflorescence 1–3-flowered. Bracts ovate, 0.6–1.6 × 0.5–1.2 cm. Flowers white, tinged greenish or partly pale pinkish, sometimes spotted. Dorsal sepal practically reduced to an erect, slender elongate spur 1–2.2 cm long mostly greenish, the petals united to the margins; lateral sepals obliquely obovate, 6.5–15 × 3–10 mm, united to about the middle, often purple-spotted, bearing conical sacs 1–1.5 mm long. Petals falcate, oblong, ribbed inside. Lip with elongate linear claw 0.8–1.5 cm long not reaching apex of spur.

Habitat and distribution In leaf litter in evergreen and bamboo forest, also in secondary forest between 1100 and 2700 m in Tigray, Gonder, Gojam, Shewa and Bale. Also in Nigeria, DR Congo, Sudan, Uganda, Kenya, Tanzania, Malawi, Zambia, Zimbabwe and South Africa.

Flowering period August and September.

Conservation status Rare in Ethiopia but locally common elsewhere.

Notes *D. anthoceros* differs from *D. meirax* by its dunce-cap like dorsal sepal being 10–22 mm long rather than 5–7 mm long.

D. anthoceros

2. D. crassicaulis *Rchb.f.*

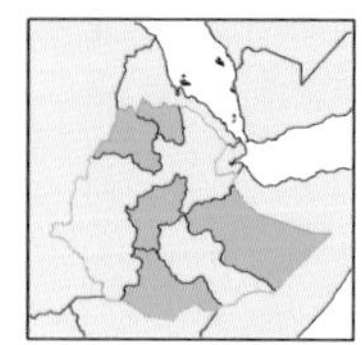

Disperis crassicaulis

The specific epithet '*crassicaulis*' refers to the relatively thick stems. H.G. Reichenbach described it in 1850 from a plant collected in Ethiopia by Georg Wilhelm Schimper.

Plant 2–5 cm tall. Leaf 1, prostrate, 10–18 × 5–12 mm, slightly cordate at the base. Bracts ovate, 5–10 × 4–7 mm. Flowers 1–2; lateral sepals pale yellow; hood pale yellow-brown to deep maroon; lip yellow-green. Dorsal sepal linear-lanceolate, 7–8 × 1 mm, joined to the petals to form an open hood. Lateral sepals ovate, oblique, 6–7 × 4 mm bearing small spur-like sacs 0.5 mm long. Petals obovate, 6 × 4 mm. Lip with a linear claw, 3 mm long.

Habitat and distribution In mountain grassland and *Podocarpus* forest. Between 2000 and 2800 m in Tigray, Gonder, Shewa, Harerge and Sidamo. Unknown elsewhere.

Flowering period May to August.

Conservation status Vulnerable.

Notes *D. crassicaulis* differs from the other Ethiopian species by having a solitary prostrate leaf rather than two or more leaves borne along the stem.

D. crassicaulis

3. D. dicerochila *Summerh.*

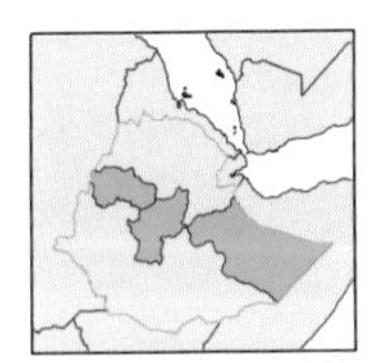

Disperis dicerochila

The specific epithet '*dicerochila*' refers to the two-horned lip. Victor Summerhayes described it from a plant collected in Ruwenzori, Uganda by William Eggeling.

Plant glabrous, 6–26 cm tall. Leaves 2, opposite. Inflorescence 1–3-flowered. Bracts lanceolate, 0.7–2.6 × 0.3–1.1 cm. Flowers white often tinged rose or purple. Dorsal sepal narrowly linear-lanceolate, 0.7–1.1 cm long, joined to the petals to form an open boat-shaped hood; lateral sepals rhomboid-ovate, 0.7–1 × 4–7.5 mm, acute, almost free, bearing obtuse sacs 1.5–2 mm long. Petals B-shaped or elliptic, 0.7–1 × 3.5–4 mm. Lip 7–9 mm long with long claw.

Habitat and distribution In leaf-litter, on mossy branches, rocks etc. in upland rain-forest; sometimes persisting in cypress plantations between 1650 and 2600 m in Gojam, Shewa and Harerge. Also in DR Congo, Uganda, Kenya, Tanzania, Malawi, Zambia and Zimbabwe.

Flowering period August to October.

Conservation status Rare and vulnerable in Ethiopia. Locally abundant elsewhere.

Notes Differs from *D. meirax* by the dorsal sepal being 10–22 mm long rather than 5–7 mm long.

D. dicerochila

4. D. galerita *Rchb.f.*

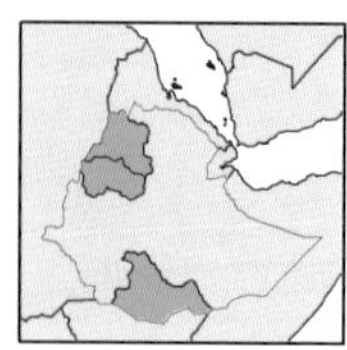
Disperis galerita

The specific epithet '*galerita*', meaning wearing a hood, refers to the hood-like dorsal sepal. H.G. Reichenbach described it in 1881 from a plant collected by Georg Wilhelm Schimper in Debre Tabor, Gonder.

Plant 10–30 cm tall. Leaves 2, rarely 3, alternate. Bracts 10–26 × 5–18 mm. Flowers 1–3, pink or purple. Dorsal sepal hooded and joined to the petals apically forming a short spur 3–4 mm long; hood 7–10 mm long. Lateral sepals obliquely ovate, 7–10 × 4–5 mm, bearing spur-like sacs 2 mm long. Petals oblong, 7–8 × 2–3.5 mm, acute, irregular on free margin. Lip 8–10 mm long; claw linear, 3–4 mm long.

Habitat and distribution In grassland, mixed grassland with *Erica arborea* forest, and *Hypericum* shrub between 2000 and 3800 m in Gonder, Gojam and Sidamo. Unknown elsewhere.

Flowering period July and August.

Conservation status Vulnerable.

Notes Differs from *D. anthoceros* and *D. meirax* by the dorsal sepal being recurved rather than erect. Similar to *D. kilimanjarica* from Uganda, Kenya and Tanzania but differs in having a shorter spur on the hood and more distinctive lateral lobes at the base of the appendage.

5. D. meirax *Rchb.f.*

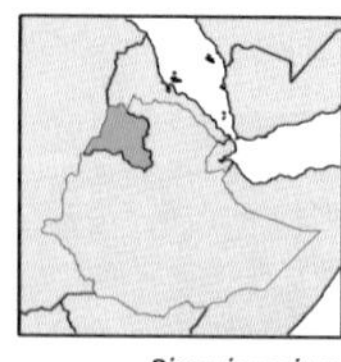
Disperis meirax

The specific epithet '*meirax*', a boy or girl, refers to the diminutive size of the plant. H.G. Reichenbach described it in 1881 from a plant collected by Georg Wilhelm Schimper in Ethiopia.

Plant 8–11 cm tall. Leaves 2, 10–22 × 5–9 mm, alternate, ovate. Bracts 6–11 × 3–6 mm. Flowers 1–2, pink to pale red. Dorsal sepal forming a conical hood, 7 × 3 mm. Lateral sepals obovate, oblique, 2.6 × 2 mm. Petals ovate, oblique, 5 × 1 mm. Lip 5 mm long; claw linear, 4 mm long.

Habitat and distribution In *Erica arborea* forest with alpine meadows on steep slopes between 3500 and 3800 m in Gonder. Unknown elsewhere.

Flowering period August and September.

Conservation status Critically endangered.

Notes *D. meirax* differs from *D. anthoceros* by the dorsal sepal being 5–7 mm long rather than 10–22 mm long.

6. D. johnstonii *Rolfe*

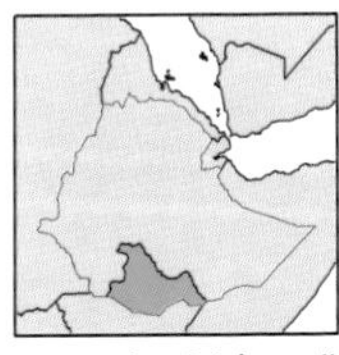
Disperis johnstonii

The specific epithet '*johnstonii*' was given in honour of the collector of the type, Sir Harry Johnston. Robert Rolfe described it in 1898 from a plant collected by Johnston on Kilimanjaro.

Plant glabrous, 4.5–15 cm tall. Leaves 2, remotely alternate, sessile, rounded-ovate to ovate-lanceolate or elliptic, 0.8–3 × 0.4–2.2 cm. Inflorescence 2–5-flowered; flowers white and pale purple or pink. Bracts broadly elliptic, 6–10 × 3.5–7 mm. Intermediate sepal linear, 0.8–1 × 0.6 mm, joined to the petals to form an open concave hood, 8–12 × 7–10 mm, sometimes with a purple-margined yellow spot on either side; lateral sepals obliquely semicircular, 8–14 × 4.5–6.5 mm, joined for about a third of their length, bearing small sacs 0.5–1 mm long. Petals narrowly elliptic, falcate, 9–10 × 2–3 mm, mostly whitish with yellow apex. Lip cream or yellowish, 4–6.5 mm long, with sharply bent claw near base.

Habitat and distribution In savannah woodland, dark brown soil at about 2000 m in the Sidamo region. Also in Nigeria, Cameroon, DR Congo, Tanzania, Zimbabwe and Malawi.

Flowering period June to September.

Conservation status Vulnerable.

Notes *D. johnstonii* differs from *D. anthoceros*, *D. galerita* and *D. meirax* by the hood being longer than broad rather than broader than long.

12. PLATYLEPIS *Rich.*

Terrestrial herbs; stem creeping at base, with tomentose roots. Leaves ovate, petiolate, sheathing at the base. Inflorescence terminal; scape with several sheaths; raceme few–many-flowered, short or long, narrow or wide; bracts conspicuous, often broad and glandular-pilose, longer than the ovaries. Sepals free; petals connivent with the dorsal sepal. Lip erect, adnate to the column for part of its length, saccate at the base with variously shaped calli, and reflexed at the apex. Column elongate; rostellum erect, bilobed; clinandrium and anther oblong, erect behind the rostellum; stigma papillose. Capsules oblong.

A genus of 10 species in tropical Africa, South Africa, Madagascar and the Mascarenes. Represented by a single species in Ethiopia.

P. glandulosa (*Lindl.*) *Rchb.f.*

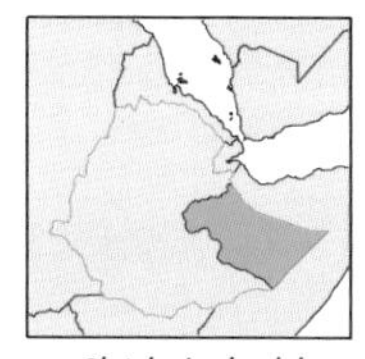

Platylepis glandulosa

The specific epithet '*glandulosa*' refers to the glandular hairs on the inflorescence. John Lindley described it as *Notiophrys glandulosa* in 1862 from plants collected by Charles Barter in Principe. H.G. Reichenbach transferred it to *Platylepis* in 1877.

Plant 15–50 cm tall; stem creeping at the base. Leaves aggregated on the lower part of the stem, lamina ovate and slightly oblique, 4–16 × 2–5.5 cm. Scape with several acuminate sheaths. Inflorescence densely many-flowered, 5–12 cm long. Bracts broad, ovate, glandular-pilose. Flowers shortly pedicellate. Sepals pale pinkish brown or yellowish green, sparsely pilose on outer surfaces, 7–9 × 1.4–2.7 mm broad; dorsal sepal slightly broader than the lateral sepals, ovate-oblong. Petals white, connivent with the dorsal sepal to form a hood, spathulate, 7.5–8.5 × 1 mm. Lip white, adnate to the column for half its length, 6.5–8.5 × 2.2–2.5 mm, bigibbous at the base with 2 rounded calli and 2 parallel linear swellings beneath the free part of the lip.

Habitat and distribution In shady, marshy places on river banks and swamps between 1000 and 1500 m in Harerge. Also throughout tropical Africa south to South Africa (Natal).

Flowering period October and November.

Conservation status Rare in Ethiopia. Widespread but always uncommon elsewhere.

P. glandulosa

13. CHEIROSTYLIS *Blume*

Small terrestrial herbs. Stems erect, arising from fleshy rhizomes. Leaves petiolate, with a sheathing base. Inflorescence terminal; with short, pubescent, up to 20-flowered raceme. Flowers small, white. Sepals joined for half their length. Petals equalling the sepals and adnate to the dorsal sepal. Lip equal to or longer than the other petals and sepals and joined to the base of the column, with 2 calli, lobed at the apex. Column short, with 2 apical appendages parallel to the elongate rostellum and 2 lateral stigmas; anther dorsal; pollinia 2, sectile; caudicle short; viscidium oblong. Capsule obovoid or oblong.

A genus of about 22 species found mainly in tropical Asia across to Australasia, with only 3 species in Africa, (one of which extends to Madagascar and the Comores). Represented by a single species in Ethiopia.

C. lepida (*Rchb.f.*) *Rolfe*

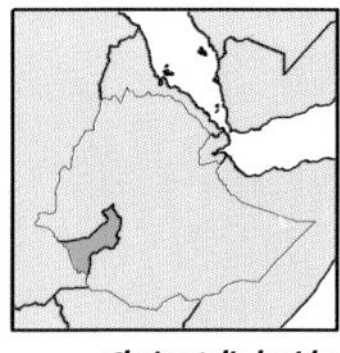
Cheirostylis lepida

The specific epithet comes from the Greek '*lepida*', meaning charming or elegant, in reference to the habit of the plant. H.G. Reichenbach described it as *Monochilus lepidus* in 1881 from a plant collected by Gustav Mann on Mt Cameroon, Cameroon. Robert Rolfe transferred it to *Cheirostylis* in 1897.

Plant usually about 15 cm tall, but sometimes up to 30 cm. Stem creeping at base. Leaves ovate, shortly acuminate, rounded or slightly cordate at the base, up to 5 × 2.5 cm; petiole shorter than the lamina, widening at the base to sheath the stem. Inflorescence terminal, sparsely pilose, with up to 20 white flowers. Bracts glabrous, exceeding ovary. Sepals and petals 3.5–4.8 × 1.1–2.7 mm. Sepals joined for half their length. Petals adnate to the dorsal sepal. Lip up to 5.6 mm long, equalling or exceeding the other petals and sepals, saccate at the base with 2 hooked calli, bilobed at the apex; lobes broad, divergent, entire or slightly toothed. Column up to 2 mm long.

Habitat and distribution In dense shade in leaf mould on forest floor between 900 and 2450 m in Kefa. Also in Nigeria, Cameroon, São Tomé, DR Congo, Rwanda, Tanzania, Kenya and Uganda.

Flowering period December to January.

Conservation status Rare and endangered in Ethiopia. Locally common elsewhere.

C. lepida

14. CORYMBORKIS *Thouars*

Terrestrial herbs. Stems erect, woody. Leaves distichous but often appearing spiral, sessile to shortly petiolate, plicate, lanceolate to elliptic. Inflorescences with few–many-flowered panicles. Flowers not opening widely, white to greenish white. Sepals and petals linear, subequal, basally connivent. Lip similar but broadly ovate at apex. Column long, slender, straight, dilated at apex with 2 lateral auricles; anther erect, more or less as long as column; pollinia 2, narrow, sectile, on a long slender stalk attached to a peltate viscidium descending behind the column; stigma broad, deeply 2-lobed; rostellum erect, bifid. Capsule retaining remnants of perianth and column.

A genus of seven pantropical species, two in Africa, one of these in Madagascar and the Mascarene Islands, probably two in Asia and three in South America. Represented by a single species in Ethiopia.

C. corymbis *Thouars*

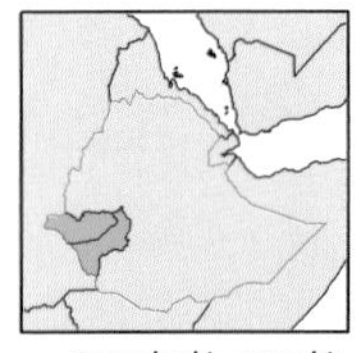
Corymborkis corymbis

The specific epithet '*corymbis*' refers to the corymb-like inflorescence. The French botanist and collector Louis Marie Aubert du Petit Thouars described it in 1822 from a plant he collected on Réunion.

Plants often forming colonies. Stems semi-woody, 0.5–2 m tall. Leaves elliptic-lanceolate, 11–35 × 3–10 cm, with a short petiole. Inflorescences 1–4 with erect to drooping panicles which are more or less one sided, up to 9 cm long, 10–16-flowered. Bracts small. Flowers white or greenish white, turning creamy white with age. Sepals and petals linear-spathulate, 45–90 × 2–5 mm. Lip clawed, 45–90 × 5–13 mm. Capsule up to 3 cm long, extended by persistent remains of column.

Habitat and distribution In evergreen forest in deep shade between 550 and 1400 m in Kefa and Illubabor. Also throughout tropical Africa and Madagascar.

Flowering period June, July; October to December.

Conservation status Rare and endangered in Ethiopia. Locally common elsewhere.

C. corymbis

15. EPIPACTIS *Zinn*

Terrestrial, occasionally saprophytic herbs with horizontal or vertical very short rhizomes, numerous fleshy roots and simple, erect, leafy stems. Leaves ovate or lanceolate, plicate, occasionally very small. Flowers in more or less secund racemes, rather inconspicuous, spreading or pendulous, shortly pedicellate. Tepals spreading or remaining closed, dull reddish or greenish. Sepals free, subequal. Petals scarcely smaller than sepals. Lip usually in 2 parts articulated by a narrow joint or fold (mesochile); the basal part (hypochile) forms a nectar-containing cup, often with a pair of basal bosses; the apical part (epichile) forms a more or less cordate or triangular downwardly directed terminal lobe; spur absent. Column short, flat or concave in front with a shallow cup at apex; anther free, hinged at the back of the summit of the column, behind the stigma and rostellum, ovate, slightly convex, 2-celled; pollinia 2, tapering towards their apices near which they are attached to the rostellum, each more or less divided longitudinally into halves; caudicles absent; pollen grains forming friable masses loosely bound by fine threads; stigma prominent, broad; rostellum placed centrally above the stigma, large, globular, persistent, evanescent or absent. Capsule oblong, spreading or pendulous.

A genus of about 25 species mainly in north temperate regions but with three species in tropical Africa. Represented by two species in Ethiopia.

Key

1 Epichile of lip broadly ovate, acute, margins forming a suberect callus on either side of mid-point; callus globular, central on epichile, extended at base in a series of warts **1. E. africana**

\- Epichile of lip fleshy, obscurely 3-lobed, with a central, raised longitudinal ridge and suberect side lobes **2. E. veratrifolia**

1. E. africana *Rendle*

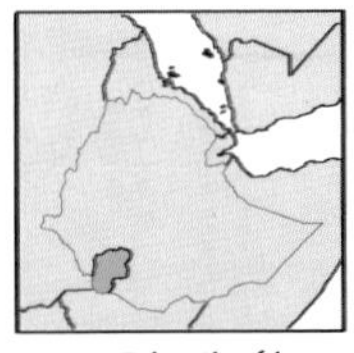
Epipactis africana

The specific epithet '*africana*' refers to Africa, where the species is endemic. Alfred Rendle described it in 1895 from a plant collected by George Scott-Elliot on Butahu, Ruwenzori, DR Congo.

Plant up to 3.5 m tall, ferruginous-pubescent in most parts. Stem purple-green below, deep blue-green

E. africana

above. Leaves lanceolate to ovate, up to 20 × 9 cm. Inflorescence elongate, few–several-flowered. Bracts 2–11 × 2 cm. Flowers rather inconspicuous, subcampanulate; sepals greyish purple on outer surface, greenish yellow within, pubescent; petals greenish yellow, purple-tinged on veins and margins; lip yellowish brown; pedicel up to 2 cm long when ovary matures. Dorsal sepal lanceolate, 3-nerved, 2 × 0.7 cm; lateral sepals broadly ovate, up to 2.4 × 0.8 cm. Petals glabrous, lanceolate to ovate, up to 2.5 × 0.6 cm, with a warty keeled mid-vein on outer surface. Lip up to 2.5 cm long; hypochile narrowly oblong, 1.2 × 0.3 cm, warty within; side lobes attached at base, as long as hypochile, linear-oblong, membranous; epichile broadly ovate, acute, 1.3 cm long, margins forming a suberect callus on either side of mid-point; callus globular, central on epichile, extended to base in a series of warts.

Habitat and distribution Grows amongst *Pteridium*, *Erica* or in coarse grass patches by rivers and in montane evergreen and bamboo forests between 2350 and 3750 m in Gamo Gofa. Also in DR Congo, Uganda, Kenya, Tanzania, Malawi.

Flowering period January.

Conservation status Rare in Ethiopia but widespread elsewhere.

Notes *E. africana* differs from *E. veratrifolia* by the epichile being broadly ovate and acute rather than fleshy and obscurely 3-lobed.

2. E. veratrifolia *Boiss. & Hohen.*

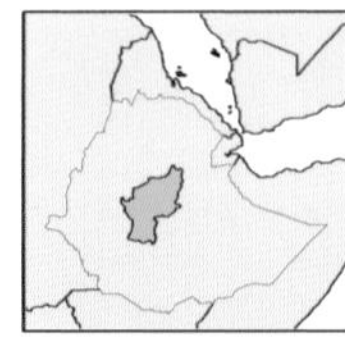
Epipactis veratrifolia

The specific epithet '*veratifolia*' refers to the *Veratrum*-like leaves. Pierre Boissier and Rudolph Hohenacker described it in 1854 from a plant collected in Iran by Theodore Kotschy. It was also known by the name *E. abyssinica* in Ethiopia.

An erect to scrambling terrestrial herb, 20–100 cm tall. Stem 8–20-leaved. Leaves 10–28 × 3–6 cm, plicate, linear-lanceolate. Inflorescence 9–24 cm long, laxly many-flowered; rhachis pubescent. Bracts 5–15 × 1–1.5 cm, lanceolate to narrowly elliptic, pubescent. Flowers with green, buff or yellow tepals, with broad marginal purple bands; lip white or buff with a purplish hypochile and a purple band on the epichile; pedicel, ovary and outer surface of flower densely pubescent. Dorsal sepal 10–19 × 4–7 mm, lanceolate; lateral sepals 10–20 × 4–8 mm, obliquely ovate. Petals 8–18 × 4–8 mm, ovate. Lip bipartite, 9–21 × 7–12 mm; hypochile up to 1 cm long, narrowly saccate, with erect sides and auriculate at base; epichile fleshy, obscurely 3-lobed, with a central, raised longitudinal ridge and suberect side lobes.

Habitat and distribution In moist areas below cliffs and on grassy hillsides or near springs and irrigation canals between 450 and 2600 m in Shewa. Also in Cyprus and Turkey east to the Himalayas and south to Yemen, Oman and Somalia.

Flowering period November and December.

Conservation status Rare in Ethiopia. Widespread but never common in south-eastern Europe, Arabia and temperate Asia.

Notes Differs from *E. africana* by the epichile being fleshy and obscurely 3-lobed rather than broadly ovate and acute.

E. veratrifolia

16. NERVILIA *Gaudich.*

Terrestrial herbs arising from small tubers. Tubers ovoid or ellipsoid, more or less dorsi-ventrally flattened, more or less pubescent. Hysteranthérous, with the leaf appearing after the flowering stem has withered. Leaf erect, suberect or prostrate, solitary, lanceolate, ovate, cordate or reniform, glabrous or pubescent; petiole short or long, often sulcate, subtended at the base by a sheathing, acute or obtuse cataphyll. Inflorescence erect, 1–many-flowered, racemose. Scape bearing 3–4 more or less sheathing cataphylls, more or less elongating after fertilisation. Flowers erect, horizontal or pendent (mostly pendent when fertilised), green, yellow, brown, white, pink or purple; bracts lanceolate, very acute; pedicel thin, ridged; ovary ellipsoidal, 3- or 6-ridged. Sepals and petals subsimilar, lanceolate. Lip entire or trilobed, more or less papillate or pubescent, sometimes spurred. Column more or less clavate; anther terminal, conical to oblong; pollinia 2, granular; stigma ventral, orbicular to triangular, towards apex of column, separated from anther by a broad blunt rostellum.

A genus of about 80 species, extending from Togo and Ethiopia through tropical Africa to South Africa, Yemen, Oman, Madagascar, Mascarenes, India and SE Asia to Japan, the Pacific islands and Australia. Three species are found in Ethiopia.

Key

1	Inflorescence 1-flowered, elongating after pollination; leaves pubescent	**3. N. crociformis**
–	Inflorescence 2-many-flowered, not elongating after pollination; leaves glabrous	2
2	Tepals 22–40 mm long; leaves very large, up to 13 × 22 cm, more or less 20-veined	**1. N. bicarinata**
–	Tepals less than 22 mm long; leaves up to 13 × 10 cm, often pleated along veins	3
3	Leaf ovate, erect, petiole more than 6 cm long	**2b. N. kotschyi** var. **purpurata**
–	Leaf cordate or broadly cordate (rarely ovate), petiole less than 6 cm long	**2a. N. kotschyi** var. **kotschyi**

N. bicarinata

1. N. bicarinata *(Blume) Schltr.*

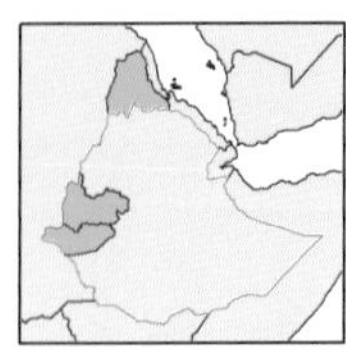
Nervilia bicarinata

The specific epithet '*bicarinata*' refers to the two keels on the lip. Carl Blume described it as *Pogonia bicarinata* in 1859 from a plant collected by August Perville in Sambirano, Madagascar. Rudolf Schlechter transferred it to *Nervilia* in 1911. Also known by the name *Nervilia ghindiana* (Fiori) Cufod.

A terrestrial erect herb to 60 cm tall. Leaves more or less 20-veined, reniform, deeply cordate at base, dark green above, more or less purple below, up to 13 × 22 cm. Inflorescence erect, laxly 4–12-flowered, flowers spaced evenly along rhachis at intervals of 2–4 cm. Bracts filamentous to lanceolate, slightly boat-shaped, to 1.7 cm long. Flowers greenish with a paler white or greenish lip, veined purple or green. Dorsal sepal ligulate-lanceolate, 20–30 × 1–4 mm. Lateral sepals slightly falcate, similar to dorsal sepal. Petals ligulate-lanceolate, 17–26 × 1–4 mm long, slightly falcate. Lip ovate, obscurely trilobed just below middle, up to 28 × 20 mm, bearing 2 more or less parallel fleshy pubescent ridges from base to midlobe; side lobes oblong to very shortly triangular; midlobe much longer than side lobes, triangular to ovate.

Habitat and distribution In primary and mixed secondary forest between 1190 and 1900 m in Illubabor and Wellega and in Eritrea. Also in Yemen and Oman, throughout tropical Africa from Senegal and Sierra Leone across to Kenya, Tanzania and south to South Africa, Comores, Madagascar, the Mascarene Islands.

Flowering period March to May. Leaves appear in July to September.

Conservation status Rare and vulnerable.

Notes Differs from *N. kotschyi* in having a large cordate leaf on a long stalk and by having tepals more than 22 mm long.

2. N. kotschyi *(Rchb.f.) Schltr.*

The specific epithet '*kotschyi*' was given in honour of Theodore Kotschy who collected a large number of plant specimens in Sudan. H.G. Reichenbach described it as *Pogonia kotschyi* in 1864 from plants collected in Sudan by Cienowski. Rudolf Schlechter transferred it to *Nervilia* in 1911.

2a. var. **kotschyi**

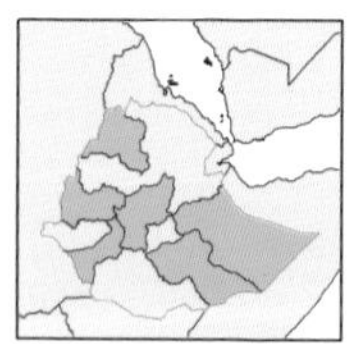

Nervilia kotschyi var. *kotschyi*

Leaf solitary, usually produced after the inflorescence has set seed, horizontal to erect, broadly cordate or more rarely broadly ovate, 3–7.5 × 4–13 cm, dark green above, sometimes purple beneath, veins often bearing a raised more or less raggedly lacerate keel. Inflorescence erect, 2–7-flowered, up to 28 cm long. Bracts linear, setose, up to 1.5 cm long. Flowers pale to olive-green with a white or off-white lip lined on veins with purple. Sepals and petals linear-lanceolate, 1.4–1.9 × 0.3 cm. Lip porrect, obscurely 3-lobed in apical half, elliptic in outline, 1.4–1.8 × 0.9–1.2 cm; side lobes shortly triangular; midlobe ovate-triangular or triangular, 0.6 cm long; callus of 2 longitudinal fleshy ridges, pubescent within.

Habitat and distribution In poor sandy and limey soils in grassland, rocky ridges in *Calpurnia aurea* scrub and in shady places in deciduous *Combretum-Terminalia* woodland between 1280 and 2200 m in Gonder, Shewa, Harerge, Bale, Kefa and Wellega. Also across to Sudan, Uganda, Kenya and Tanzania and south to Zambia and Zimbabwe.

Flowering period April to May. Leaves appear between June and August.

Conservation status Vulnerable.

Notes Differs from *N. bicarinata* in having a smaller cordate leaf on a short stem and much small flowers with the tepals less than 22 mm long.
Var. *kotschyi* differs from var. *purpurata* in its leaf shape and by the petiole being less than 6 cm long rather than more than 6 cm long.

N. kotschyi var. *kotschyi*

2b. var. purpurata (*Rchb.f. & Sond.*) *B. Pettersson*

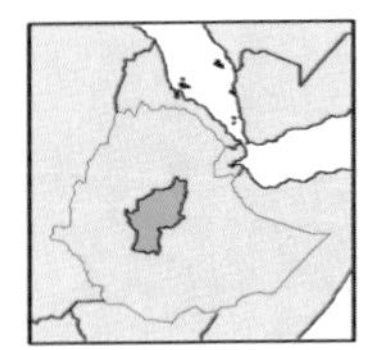

Nervilia kotschyi var. *purpurata*

The specific epithet '*purpurata*' refers to the purple underside of the leaf. H.G. Reichenbach and Otto Sonder described it as *Pogonia purpurata* in 1865 from a plant collected by Carl Zeyher in S. Africa. Börje Pettersson transferred it to *N. kotschyi* var. *purpurata* in 1990.

Plant up to 22 cm high. Leaf solitary, appearing after the inflorescence has withered, erect, ovate or elliptic, up to 10 × 4 cm, green above, more or less purple beneath, prominently veined, borne on a more or less elongate slender petiole up to 13 cm or more long. Inflorescence 2–4-flowered, up to 22 cm long. Bracts linear-subulate, up to 2 cm long. Flowers with green or yellow-green sepals and petals and pale green or yellowish lip lined with purple-brown on veins. Sepals and petals linear-lanceolate, 1.8–2.2 × 0.4 cm. Lip obscurely 3-lobed, elliptic-oblong in outline, 1.8 × 1.3 cm, pubescent between the 2 central longitudinal fleshy ridges; side lobes erect, narrowly oblong, truncate; midlobe triangular-ovate.

Habitat and distribution In grassland between 475 and 2300 m in Shewa. Also in Tanzania, DR Congo, Zambia and Angola from south to South Africa.

Flowering period Unknown.

Conservation status Endangered.

Notes Var. *purpurata* is distinguished from var. *kotschyi* by its ovate erect leaf borne on a petiole more than 6 cm long.

Leaves of *N. kotschyi* var. *purpurata* and *N. crociformis* (centre)

3. N. crociformis (*Zoll. & Mor.*) *Seidenf.*

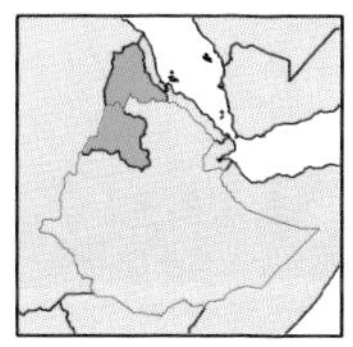

Nervilia crociformis

The specific epithet '*crociformis*' refers to the crocus-like appearance of the flower before the leaf appears. Heinrich Zollinger and Alexandre Moritzi described it as *Bolborchis crociformis* in 1846 from a plant Zollinger collected in Java. Gunnar Seidenfaden transferred it to *Nervilia* in 1978.

Plant up to 10 cm tall. Leaf appressed to ground, reniform, minutely apiculate, very cordate at base, densely covered above with short white hairs, 1.5–4 × 3.5–7 cm. Scape terete, 1-flowered at apex, with 2–3 clasping sheathing cataphylls. Bract linear, equal to pedicel in length. Flower brownish green; lip white with a yellow centre. Sepals spreading, linear-ligulate, 12–19 × 3–3.5 mm; lateral sepals oblique. Petals similar to sepals but shorter. Lip oblong-cuneate, 3-lobed in apical third, 1.3 × 0.8 cm, surface covered with scattered long hairs, base finely 3-keeled, with mid-keel muricate-mammillate; lateral lobes triangular; midlobe a little longer, ovate-triangular to ovate, margin crenulate-undulate.

Habitat and distribution In grassland and forest, in rich soil between 1200 and 2200 m in Gonder and in Eritrea.Widespread in tropical Africa and Madagascar.

Flowering period June and July.

Conservation status Locally rare and endangered in Ethiopia. Common elsewhere.

Notes Distinguished from *N. bicarinata* and *N. kotschyi* by the 1-flowered inflorescence and pubescent leaves.

N. crociformis

17. MALAXIS *Sw.*

Small terrestrial lithophytic or epiphytic herbs arising from a creeping rhizome. Secondary stems erect, leafy, more or less swollen at base, clustered or not, 1–several-leaved. Leaves thin-textured, plicate, mostly ovate or broadly ovate, with sheathing more or less elongate leaf-bases. Inflorescence terminal, erect, mostly densely many-flowered, racemose to subumbellate. Flowers mostly small, non-resupinate, green, buff, orange, or purple. Sepals and petals subsimilar or petals filiform, spreading. Lip larger than other floral segments, flat, entire or lobed, more or less auriculate at base on either side of column, margins more or less toothed. Column short, porrect, more or less auriculate; stigma ventral; anther terminal, flap-like; pollinia ovoid, 4 in 2 pairs joined at base.

A large genus of about 300 species of cosmopolitan distribution, but predominantly tropical Asian. About seven species known from Africa. Represented by a single species in Ethiopia.

M. weberbaueriana *(Kraenzl.) Summerh.*

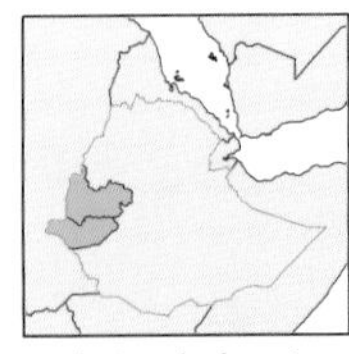
Malaxis weberbaueriana

The specific epithet '*weberbaueriana*' was given in honour of Otto Weberbauer who collected the type specimen from Mt Cameroon, Cameroon. Fritz Kraenzlin described it as *Microstylis weberbaueriana* in 1908. Victor Summerhayes transferred it to *Malaxis* in 1934.

Plant 10–25 cm tall, often growing in large groups. Secondary stems erect, 2.5–8 cm tall, leafy near apex, 3–5-leaved. Leaves ovate or elliptic-ovate, up to 5.5 × 3 cm. Inflorescence laxly many-flowered, up to 17 cm long. Bracts ovate, 3–4 mm long. Flowers purple or less commonly green. Dorsal sepal ovate, 2.3–3.3 × 1.7 cm. Lateral sepals obliquely ovate, 2–3 × 1.6 mm, shortly united at base. Petals obliquely linear-lanceolate, 2.2–3 × 0.6 mm. Lip subcircular to subquadrate, 2 × 2 mm, bearing two lateral lunate pubescent calli.

Habitat and distribution In humus in riverine and montane forest between 1500 and 1800 m in Illubabor and Wellega. Also in Cameroon, Fernando Po, Kenya, Tanzania, Malawi, Zambia and Zimbabwe.

Flowering period June and July.

Conservation status Rare in Ethiopia but widespread and locally common elsewhere.

M. weberbaueriana

18. LIPARIS *Rich.*

Terrestrial and epiphytic herbs. Stems usually more or less swollen at the base to form pseudobulbs. Leaves basal or sometimes cauline, either broad, plicate, thin in texture and unjointed at the base, or else small, narrow, firm-textured and jointed at the base. Inflorescences few–many-flowered, terminal, mostly erect; rhachis terete or sometimes flattened or angular. Flowers rather small, usually yellowish green or purplish. Tepals spreading or reflexed, narrow, mostly with entire margins; petals often linear. Lip usually more fleshy than the tepals, simple or bilobed, with entire, dentate or crenulate margins, often bent in the middle, with 2 simple calli at the base. Column relatively elongated, usually curved over, terete or slightly winged, with or without a small foot; pollinia 4, pear-shaped, in 2 pairs.

A genus of about 250 species, mainly in the tropics, but also in Europe, North America, Australia and New Zealand. twelve species are recorded for East Africa. Three species are found in Ethiopia.

Key

1	Pseudobulbs well separated on a long rhizome; lateral sepals united almost to apex	**2. L. deistelii**
–	Pseudobulbs clustered or solitary; lateral sepals free to middle or base	2
2	Pseudobulbs subterranean; leaves lacking prominent pleats; lateral sepals united in basal half; lip subrhombic	**1. L. abyssinica**
–	Pseudobulbs above the ground; leaves with prominent pleats; lateral sepals free to base; lip obovate	**3. L. nervosa**

1. L. abyssinica *A.Rich.*

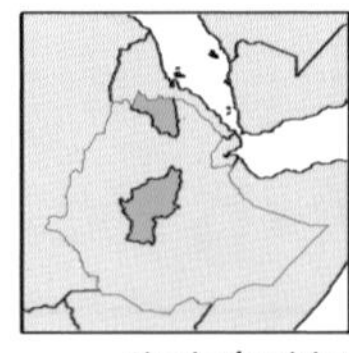

Liparis abyssinica

The specific epithet '*abyssinica*' refers to Abyssinia, the older name for the present-day Ethiopia. Achille Richard described it on 1851 from plants collected by Richard Quartin-Dillon in Mt Selloda, near Adwa.

Plant 8–12 cm tall. Pseudobulbs subterranean, conical to subcylindrical, up to 2 cm long, borne at an angle to the previous year's growth, 2–5-leaved. Leaves lanceolate to ovate or broadly ovate, acute, 2–7 × 0.5–1.8 cm, lacking prominent longitudinal pleats. Inflorescence up to 11 cm long, few-flowered, petiole clasping at the base; peduncle bearing 1–several sterile

L. deistelii

bracts; bracts narrowly lanceolate, acuminate, 5–10 × 1 mm. Dorsal sepal oblong-lanceolate, acute, 6 × 2 mm. Lateral sepals adnate in basal half, falcate, narrowly elliptic, obtuse, 4.5 × 2 mm. Petals deflexed-spreading, linear, rounded at apex, 5.5 × 0.8 mm. Lip strongly reflexed, subrhombic, obtuse or rounded at apex, 5–7 × 6–8 mm; callus near base of lip small, somewhat bifid, truncate in front. Column more or less incurved at right angle in middle, 3 mm long.

Habitat and distribution The species is found in montane forest and grassland between 2000 and 2300 m in Tigray and Shewa. Unknown elsewhere.

Flowering period August.

Conservation status Critically endangered. Known only from the type and two other collections.

Notes Distinguished from *L. deistelii* by its terrestrial habit and the lateral sepals being united in the basal half only rather than almost to the apex.

2. L. deistelii *Schltr.*

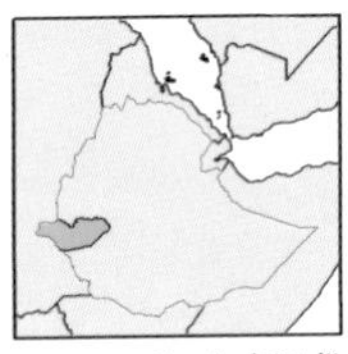

Liparis deistelii

The specific epithet was given in honour of Hans Deistel, the collector of the type specimen. Rudolf Schlechter described it in 1906 from a plant collected by Deistel above Buea, Mt Cameroon, Cameroon.

Plant to 15 cm tall, arising at intervals of 4–12 cm from a slender terete creeping rhizome. Pseudobulbs fleshy, up to 4 × 4 mm, rarely broader, with 3 larger foliage leaves arising at intervals up the pseudobulb. Leaves ovate to oblong-lanceolate 5–12 × 1.5–3 cm, 5-veined, margins smooth or slightly crinkled. Inflorescence terminal, longer or shorter than the leaves, 2–10-flowered; rhachis up to 3 cm long; peduncle up to 9 × 1 mm. Bracts up to 5 × 2 mm. Ovaries slightly longer than the bracts. Flowers cream or yellowish green to light and dark reddish purple. Dorsal sepal up to 10 × 3 mm at the auriculate base; lateral sepals joined almost to the apex although they may pull apart slightly; united part up to 6.5 × 6.5 mm, orbicular, slightly auriculate at the base, curved up beneath the lip. Petals linear, up to 11.5 × 0.8 mm. Lip narrow at the auriculate base with a bilobed callus, then abruptly widening to become fan-shaped, emarginate, apiculate, margins crenulate to dentate, up to 7.5 × 7 mm.

Habitat and distribution Epiphyte on second storey trees in montane forest, frequently with mosses and ferns between 1700 and 2750 m in Illubabor. Also in Cameroon, Gabon, DR Congo, Uganda, Kenya, Tanzania and Malawi.

Flowering period June to November.

Conservation status Rare and vulnerable.

Notes Differs from *L. abyssinica* by its epiphytic habit, elongated rhizome and the lateral sepals being united almost to apex, rather than in the basal half only.

L. deistelii

3. L. nervosa (*Thunb.*) *Lindl.*

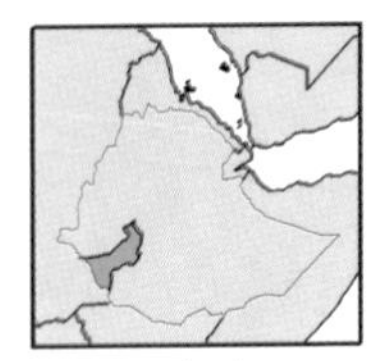

Liparis nervosa

Originally described in 1830 as *Orchis nervosa* by the Swedish botanist Carl Peter Thunberg from a plant he collected in Japan. John Lindley transferred it to the present genus in 1830 in his *Genera and Species of Orchidaceous Plants*.

A terrestrial or lithophyte 20–55 cm tall. Pseudobulbs ovoid, up to 3.5 cm long. Leaves 2–3, lanceolate to ovate, 12–35 × 2.5–5 cm, 5-veined, boldly pleated. Inflorescence terminal, longer than the leaves, densely many-flowered; rhachis and peduncle ridged. Flowers yellowish green with a yellow-green or dark maroon purple lip. Dorsal sepal 5–6.5 × 1.5–2 mm; lateral sepals often rolled up under lip, free to the apex, 3.5–5 × 1.5–2 mm. Petals linear, 4.5–6 × 0.8 mm. Lip obovate, strongly recurved, 2.5–4 × 2.5–4 mm, with a basal bilobed callus. Column slightly incurved, 2.5–3 mm long.

Habitat and distribution Terrestrial in marshes or lithophytic in forest between 1300 and 1500 m in Kefa. Also throughout tropical Africa as far south as Zimbabwe and Angola, in tropical and subtropical Asia and the tropical Americas.

Flowering period June.

Conservation status Rare and vulnerable.

Notes Differs from *L. abyssinica* by its strongly pleated leaves and the lateral sepals being free.

L. nervosa

19. OBERONIA *Lindl.*

Small epiphytic herbs with few–many clustered stems and wiry roots at base of stem. Stems leafy throughout. Leaves bilaterally flattened, iridiform, usually decreasing in size towards apex of stem. Inflorescence terminal, cylindrical, densely many-flowered, flowers in whorls; bracts minute. Flowers non-resupinate, minute, flat, translucent. Sepals and petals free, spreading. Lip flat, lacking a callus, usually with a shallow glistening basal depression. Column fleshy, very short; pollinia 4, waxy.

A genus of about 100 species, the majority in tropical Asia across to the SW Pacific Islands. A single species in tropical Africa and Madagascar which is also found in Ethiopia.

O. disticha *(Lam.) Schltr.*

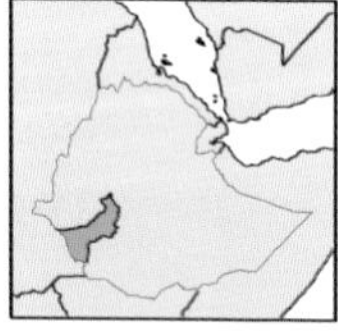

Oberonia disticha

The specific epithet refers to the distichous leaf arrangement. Lamarck described it as *Epidendrum distichum* based on a collection from Réunion. Rudolf Schlechter transferred it to *Oberonia* in 1924.

Stems leafy throughout, up to 15 cm long. Leaves bilaterally flattened, iridiform, 2–5 × 0.6–1 cm, decreasing in size towards apex of stem. Inflorescence terminal, cylindrical, densely many-flowered; rhachis papillate on ribs; bracts up to 1.5 mm long. Flowers non-resupinate, minute, flat, fleshy, yellow to orange. Dorsal sepal concave, ovate, obtuse, 0.5–0.6 mm long, papillate on outer side at apex. Lateral sepals similar but oblique. Petals elliptic-oblong, rounded or obtuse at apex, 0.5 mm long. Lip cucullate, pandurate, obtuse, 1 × 0.6–0.8 mm. Column fleshy, very short.

Habitat and distribution Epiphytic on bushes in coffee plantation at 1700 m. Known only from Kefa in Ethiopia. Widespread in tropical Africa from Cameroon across to Kenya and south to Zimbabwe, and also in Madagascar and the neighbouring islands.

Flowering period January.

Conservation status Rare in Ethiopia. Locally common elsewhere.

Notes Readily recognized by its iridiform leaves with overlapping bases borne on an elongate stem and by the cylindrical inflorescence of minute flowers.

O. disticha

20. POLYSTACHYA *Hook.*

Small, medium-sized or rarely large epiphytic or less commonly lithophytic or terrestrial herbs. Stems often pseudobulbous, caespitose or less commonly spaced on a creeping rhizome, sometimes branched or superposed, 1–several-noded, 1–several-leaved. Leaves suberect to spreading, thin-textured, coriaceous or rarely fleshy, often distichous, linear or lanceolate to oblong-elliptic or oblanceolate, emarginate, acute, obtuse or acuminate at the apex. Inflorescence terminal, erect to pendulous, 1–many-flowered, simple or branching; branches sometimes secund; bracts suberect to reflexed, setose or lanceolate to ovate or obovate, acute or acuminate to mucronate. Flowers minute to fairly large, mostly with lip uppermost, non-resupinate, mostly rather drably coloured, more or less fragrant, often pubescent. Dorsal sepal mostly porrect, lanceolate to ovate-elliptic; lateral sepals more or less oblique, attached to the column-foot to form a more or less prominent mentum. Petals linear to obovate. Lip entire to 3-lobed, with or without a basal callus, glabrous, pubescent or farinose, often recurved and difficult to flatten. Column porrect, mostly short and stout, with a more or less elongate foot; pollinia 2, ovoid; stipe 1, square or subtriangular to oblong or linear; viscidium small to large, circular or elliptic; rostellum mostly obscure, bifid in front, rarely slightly elongate and beak-like.

A large genus of about 200 species, predominantly African but also found in the tropics of Central and South America, and from Madagascar across to the Philippines and the Malay archipelago. Twelve species are found in Ethiopia.

Key

1	Leaf solitary at apex of pseudobulb	**12. P. cultriformis**
–	Leaves two or more on pseudobulb or stem	2
2	Pseudobulbs or stems superposed, each new growth emerging from a node above the base of the previous one	3
–	Pseudobulbs or stems clustered, each new growth arising from base of previous growth	5
3	Flowers fairly large, mentum at least 35 mm long; stems slender, less than 2 mm in diameter; inflorescence unbranched	4
–	Flowers small, mentum up to 3 mm long; stems fusiform to narrowly cylindrical, more than 3 mm in diameter; inflorescence branching or not	**6. P. simplex**

P. bennettiana

4	Leaves more than 6.5 mm broad; lip claw glabrous, midlobe pubescent	**7. P. aethiopica**
–	Leaves less than 6 mm broad; lip claw pubescent, midlobe glabrous	**8. P. lindblomii**
5	Plants flowering when leafless	6
–	Plants flowering when in leaf	7
6	Inflorescence branched; peduncle and rhachis covered with many papery sheaths; lip with a fleshy obconical callus; mentum 2 mm long	**10. P. steudneri**
–	Inflorescence unbranched; peduncle bearing 2 papery sheaths; lip lacking a callus, covered with a farina; mentum 3–3.5 mm long	**11. P. eurychila**
7	Inflorescence branched	8
–	Inflorescence simple	10
8	Inflorescence pyramidal in outline; flowers brick red to orange with red venation; pseudobulbs bilaterally compressed	**3. P. paniculata**
–	Inflorescence with short secund branches; flowers not as above; pseudobulbs obscurely ovoid	9
9	Leaves obovate to oblanceolate, flat, thin-textured, green; flowers pale yellow, green or pink	**4. P. tessellata**
–	Leaves linear, gutter-shaped, fleshy-coriaceous, grey-green; flowers yellow	**5. P. golungensis**
10	Plants with elongate stems, never pseudobulbous, more than 15 cm tall	11
–	Plants pseudobulbous at base, less than 8 cm tall	**9. P. caduca**
11	Mentum up to 5 mm long; sepals acuminate; lip usually longer than broad when flattened	**1. P. bennettiana**
–	Mentum 5.5–7 mm long; sepals acute or shortly acuminate; lip at least as broad as long when flattened	**2. P. rivae**

1. P. bennettiana *Rchb.f.*

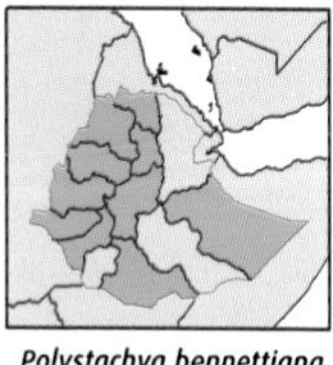
Polystachya bennettiana

The specific epithet '*bennettiana*' refers to John Bennett, Keeper of Botany at the British Museum (Natural History). H.G. Reichenbach described it in 1881 from plants collected in Ethiopia by Georg Wilhelm Schimper.

Plant 13–35 cm tall. Stems covered with loose tubular leaf-bases, 2–6-leaved. Leaves well separated, ligulate to narrowly elliptic, 10–23 × 1.2–2 cm. Inflorescence many-branched, 9–31 cm long; peduncle 1–9 cm long; rhachis terete, pubescent; branches secund, 0.5–4 cm long. Bracts ovate to ovate-lanceolate, 2–5 mm long. Flowers pubescent on outer surface, light greenish yellow or cream; lip lined with brown or red at base and on side lobes, vanilla-scented; ovary densely pubescent. Dorsal sepal oblong-lanceolate, 6.7–9.7 × 2.2–3.5 mm; lateral sepals obliquely triangular-ovate or lanceolate, acute, 9.5–11.5 × 5.5–6.7 mm; mentum broadly conical, 5.8–6.2 cm long. Petals oblanceolate, 6–8 × 1.6–2.5 mm. Lip shortly clawed, 3-lobed at or above the middle, 7.5–10 × 6–7.3 mm, clavate-pubescent all over upper surface; side lobes obliquely triangular; midlobe ovate to rotund, 3.3–5 × 2.7–3.3 mm; disc with a central pubescent fleshy keel in basal third of lip.

Habitat and distribution In open woodland, *Podocarpus* forest, secondary or riverine forest between 900 and 2430 m in Tigray, Gonder, Gojam, Welo, Shewa, Harerge, Sidamo, Kefa, Illubabor and Wellega. Also in Uganda, Kenya, Tanzania, Nigeria, Cameroon, DR Congo, Zambia.

Flowering period October, January to June.

Conservation status Widespread and not threatened.

Notes Differs from *P. rivae* by the lip being longer than broad.

2. P. rivae *Schweinf.*

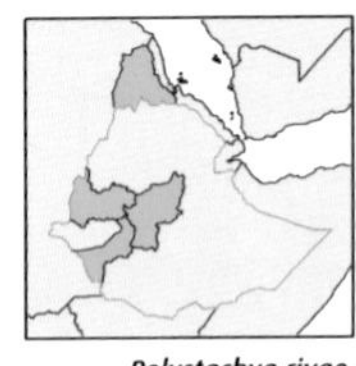
Polystachya rivae

The specific epithet '*rivae*' was given in honour of the Italian collector Domenico Riva, one of the collectors of the type. Georg Schweinfurth described it in 1894 from plants he and Riva collected in Saganeiti and Mt. Alan Kale near Aidereso, Eritrea.

Plant erect to pendulous with clustered terete stems covered by tubular sheaths, 7–30 cm long. Leaves 2–5, linear, 8–22 × 0.8–2.9 cm, glossy olive green above, prominently veined beneath. Inflorescence simple or weakly branched at base, 3–17 cm long, densely many-flowered; peduncle covered by imbricate, slightly bilaterally compressed sheaths; rhachis pubescent. Bracts lanceolate, 5–10 mm long. Flowers cream, white or greenish yellow, marked with yellow or pink on lip, sparsely pubescent on outside, scented of lily-of-the-valley; pedicel and ovary 3–7 mm long, pubescent. Dorsal sepal ovate, 6–9.5 × 3.5–4.6 mm. Lateral sepals obliquely triangular, 8–12.5 × 6.5–10 mm; mentum obliquely conical, 6–8 mm long. Petals obovate to oblanceolate, 7–8 × 1.2–2.9 mm. Lip 3-

lobed in apical half, as broad as or broader than long, 7–9 × 8–10 mm, usually clavate and farinose on inner surface and callus; side lobes almost as long as midlobe; midlobe broadly ovate-subcircular; callus fleshy, in middle of lip.

P. rivae

Habitat and distribution On trunks and limbs of small trees and gallery forest by streams and on scattered trees in grassland between 1350 and 2490 m in Shewa, Kefa and Wellega and in Eritrea. Also in Sudan.

Flowering period December, January, April, May and August.

Conservation status Vulnerable.

Notes Differs from *P. bennettiana* by the lip being as broad as or broader than long.

3. P. paniculata (*Sw.*) *Rolfe*

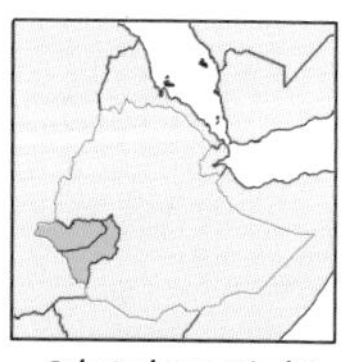

Polystachya paniculata

The specific epithet '*paniculata*' refers to the paniculate inflorescence. Olof Swartz described it as *Dendrobium paniculatum* in 1805 from plants collected by Adam Afzelius in Sierra Leone.

Plant 22–40 cm long. Pseudobulbs cylindrical, clustered, strongly compressed, 3–4-noded, 3–4-leaved. Leaves distichous, ligulate, unequally bilobed at the apex, 10–30 × 2–3.4 cm. Inflorescence generally longer

P. paniculata

than the leaves, a many-branched and many-flowered raceme, up to 21 cm long; peduncle 5–13 cm long; rhachis glabrous; branches up to 6 cm long, many-flowered. Flowers flame-red or orange with red markings on the lip. Dorsal sepal lanceolate, 3 × 0.8 mm; lateral sepals obliquely lanceolate, 3.7 × 1.8 mm, forming with the column-foot an obscurely conical mentum 1.3 mm long. Petals oblanceolate, 2.7–2.8 × 0.5 mm. Lip ovate to elliptic, entire, 2.8 × 1.7 mm.

Habitat and distribution In rainforest between 900 and 1900 m in Kefa and Illubabor. Also in Sierra Leone, Liberia, Ivory Coast, Ghana, Nigeria, Cameroon, Gabon, Congo (Brazzaville) and DR Congo.

Flowering period October to December.

Conservation status Rare and threatened in Ethiopia. Locally common elsewhere in its range.

Notes Differs from *P. tessellata* by the inflorescence being pyramidal in outline rather than secund and by its smaller orange-red flowers.

4. P. tessellata *Lindl.*

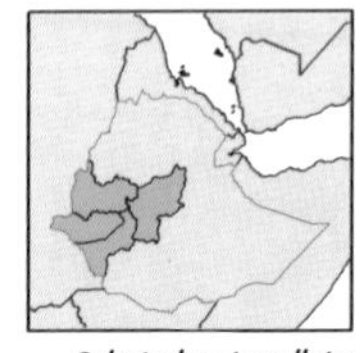

Polystachya tessellata

The specific epithet '*tessellata*' refers to the tessellated effect of the cells on the lip. John Lindley described it in 1862 from plants collected by Gustav Mann in Niger Delta at the mouth of the Nun River, Nigeria.

Plant 20–62 cm long. Stems pseudobulbous at base, to 15 × 0.7 cm, 3–5-leaved; pseudobulbs ovoid, 0.6–1.3 × 0.5–0.8 cm. Leaves distichous, oblanceolate, elliptic or oblong-elliptic, minutely unequally bilobed at apex, 13–30 × 1.3–5.7 cm. Inflorescence much longer than the leaves, paniculate, 18–47 cm long; peduncle covered with scarious imbricating sheaths; rhachis covered with papery acute sheaths, sparsely pubescent; branches secund, distant, 1–5 cm long, densely many-flowered. Bracts linear-lanceolate, 2–4 mm long. Flowers sparsely pubescent on outer surface, creamy yellow, pale green, pink or dull red, lip white with a pink midlobe. Dorsal sepal oblong-elliptic, 3–4.3 × 1.8–2.3 mm; lateral sepals obliquely ovate, 3–5.6 × 2.8–3.6 mm, forming with the column-foot a conical mentum to 3.4 mm high. Petals ligulate to spathulate, 2–3.6 × 0.6–1.1 mm. Lip clawed, 3-lobed at or above the middle, 3.5–5.2 × 2.5–4 mm; side lobes falcately oblong, 1.2 mm long; midlobe suborbicular, emarginate, tessellated when dry, margins crenulate-undulate, 1.6–2.5 × 2–2.2 mm; callus a fleshy ridge 1.7 mm long, tapering towards the base and rounded at the apex.

Habitat and distribution An epiphyte growing with mosses and ferns on trees in damp valleys, forest canopies and in riverine forest between 1100 and 1800 m in Shewa, Kefa, Illubabor and Wellega. Also throughout tropical Africa.

Flowering period August to October.

Conservation status Vulnerable in Ethiopia, but common throughout its extensive range.

Notes Differs from *P. paniculata* in flower colour and by the inflorescence being secund rather than pyramidal in outline. Some authorities sink this species into the pan-tropical *P. concreta*. It is generally a larger plant with distinctive oblanceolate leaves and we prefer to keep it distinct.

P. tessellata

5. P. golungensis *Rchb.f.*

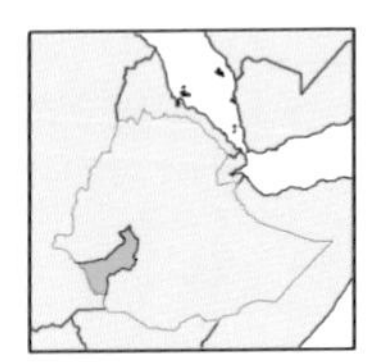

Polystachya golungenis

The specific epithet commemorates Golungo Alto in Angola where Friedrich Welwitsch collected the type material. H.G. Reichenbach described it in 1865.

Plant 11–47 cm long. Stems clustered, obscurely pseudobulbous at base, 2–6 × 0.4–0.7 cm, 2–4-leaved; pseudobulbs ovoid, 0.6–1.3 × 0.5–0.8 cm. Leaves distichous, conduplicate, fleshy, linear-ligulate, 5–17 × 0.4–2 cm, grey-green. Inflorescence much longer than the leaves, paniculate, 18–47 cm long; peduncle covered with scarious imbricating sheaths; rhachis covered with papery acute sheaths, sparsely pubescent; branches up to 1.5 cm long, secund, densely many-flowered. Bracts linear-lanceolate, 1.5 mm long. Flowers sparsely pubescent on outer surface, yellow. Dorsal sepal oblong-elliptic, 1.5–2.8 × 0.7–1.5 mm; lateral sepals obliquely ovate, 2.2–3.6 × 1.2–2.4 mm, forming with the column-foot a conical mentum to 2.2 mm long. Petals ligulate to spathulate, 1.1–2.6 × 0.3–0.61 mm. Lip clawed, 3-lobed at or above the middle, 2.2–3 × 1.7–2.5 mm; side lobes falcately oblong, 1 mm long; midlobe suborbicular, emarginate, tessellated when dry, margins crenulate-undulate, 1.3–1.4 × 1.3–1.4 mm; callus a cushion of clavate hairs at the base of the lip.

P. golungensis

Habitat and distribution An epiphyte or lithophyte growing in riverine forest, montane forest or drier woodland up to 1800 m in Kefa. Also throughout tropical Africa.

Flowering period August to October.

Conservation status Vulnerable in Ethiopia, but common throughout its extensive range.

Notes Differs from *P. concreta* in having grey-green linear fleshy leaves and rather smaller flowers with a very low mentum.

6. P. simplex *Rendle*

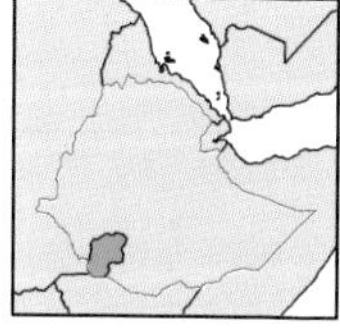

Polystachya simplex

The specific epithet '*simplex*' refers to the entire lip. Alfred Rendle described it in 1895 from plants collected by George Scott Elliot in Ruimi, Toro District, Uganda.

Plant pendulous, 50 cm or more long. Stems (or pseudobulbs) cylindrical or narrowly fusiform, 3–20 cm long, 3–9 mm wide, yellow or purple, superposed, elliptic, oblong or oblanceolate, 5–15 × 1.5 cm. Inflorescence 3.5–9 cm long, pubescent, shorter than the terminal leaves. Bracts narrowly lanceolate, 0.4–0.7 cm long. Flowers green or yellowish green, tinged with brown or mauve, lip yellow marked with purple. Dorsal sepal ovate-lanceolate, 4.8 × 2 mm; lateral sepals obliquely ovate-triangular, 5.5 × 3 mm, forming with the column-foot a mentum 2.6 mm long. Petals linear or lanceolate, 3.8 × 1 mm. Lip shortly clawed, 3-lobed above or at the middle, 5 × 3.8 mm, lacking a callus or with a slight fleshy thickening between the side lobes; side lobes erect, elliptic; midlobe ovate or rotund, 2.3 × 2.2 mm wide.

P. simplex

Habitat and distribution An epiphyte in montane forest between 1200 and 1800 m in Gamo Gofa. Also in Uganda, Kenya, Tanzania, DR Congo, Malawi and Zimbabwe.

Flowering period April.

Conservation status Rare and possibly endangered in Ethiopia but locally common elsewhere in its range.

Notes Distinguished from *P. aethiopica* and *P. lindblomii* by the mentum being less than 3 mm long.

7. P. aethiopica *P.J.Cribb*

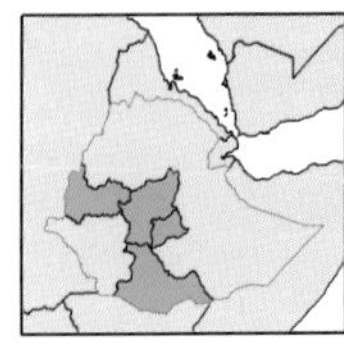

Polystachya aethiopica

The specific epithet '*aethiopica*' refers to Ethiopia, where the species is endemic. Phillip Cribb described it in 1978 from plants collected in Ethiopia by Mooney.

Stems several-noded, 2–4-leaved. Leaves narrowly oblong-ligulate, 4–8.5 cm × 6–8.5 mm, acutely and unequally bilobed at apex, prominently 5-veined below. Inflorescence erect, 7–16-flowered, 5–7 cm long. Bracts setaceous, 2–12 mm long. Flowers yellow-green; lip crimson; pedicel and ovary and outside of flower pubescent. Dorsal sepal lanceolate, 5 × 2 mm; lateral sepals very obliquely falcate-lanceolate, 5–6 × 5–6 mm at base, forming with the column-foot a curved, cylindric-conical mentum 3–4 mm long. Petals linear-lanceolate, 5 × 1 mm. Lip clawed, obscurely 3-lobed in the middle, 6.5–6.6 × 4 mm, pubescent on

P. aethiopica

Polystachya aethiopica

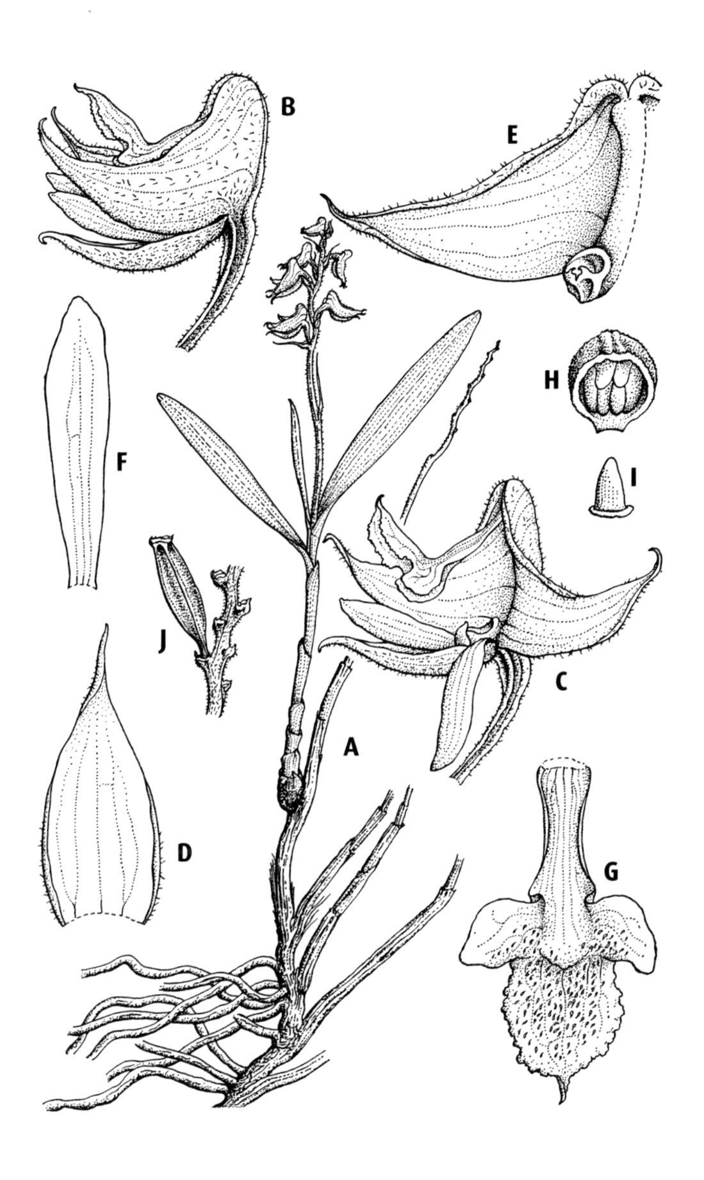

A habit, × ⅔; **B** flower, × 4; **C** flower (lateral sepal & petal turned back), × 4; **D** dorsal sepal, × 6; **E** lateral sepal with column, × 6; **F** petal, × 7; **G** labellum, × 7; **H** anther cap, × 12; **I** stipes & viscidium, × 12; **J** capsule, × 1½. Drawn from *Mooney* 7283 by Mrs. M. E. Church.

upper surface; side lobes erect, oblong-auriculate; midlobe oblong, margins erose; small fleshy callus at base of midlobe.

Habitat and distribution Amongst low shrubs between 2000 and 2800 m in Shewa, Arsi, Sidamo and Wellega. Unknown elsewhere.

Flowering period February to May.

Conservation status Rare and endangered.

Notes Differs from *P. lindblomii* by the leaves being at least 6.5 mm broad.

8. P. lindblomii *Schltr.*

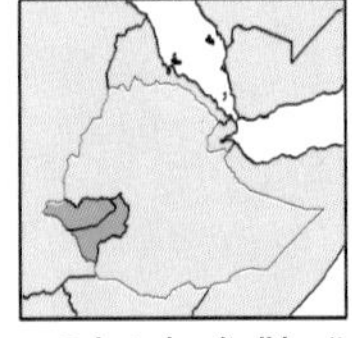

Polystachya lindblomii

The specific epithet '*lindblomii*' was given in honour of Herr Lindblom who collected the plant from Kenya. Rudolf Schlechter described it in 1922 from plants collected by Lindblom in Kitosh, N. Kavirondo District, Kenya.

A pendulous or erect plant up to 35 cm or more long. Stems up to 25 × 1 mm, secondary and tertiary stems arising from apical half of older growths, 2–9 × 0.1 cm, 5–8-leaved in apical half. Leaves linear, grass-like, flexuous, 5–11 cm long, 2–3.5 mm wide. Inflorescence unbranched, much shorter than the leaves, 4–8-flowered, 1.5–3 cm long. Bracts lanceolate, 1.5–2.5 mm long. Flowers white, greenish yellow or

P. lindblomii

yellow, with a yellow or purple lip. Dorsal sepal ovate, cucullate, 3–5 × 4–5.5 mm, forming with the column-foot a shortly and broadly cylindrical mentum 4–5.2 mm long. Petals linear to oblong-elliptic, 2.7–3.5 × 0.8–1.1 mm. Lip with a long claw, obscurely 3-lobed in apical half, 6.5 × 4 mm; side lobes erect, narrowly oblong; midlobe much larger than side lobes, broadly ovate to orbicular; claw pubescent; callus a fleshy transverse ridge between side lobes, sometimes sparsely pubescent.

Habitat and distribution An epiphyte in dense forest and at forest margins between 1000 and 1400 m in Kefa and Ilubabor. Also in DR Congo, Uganda, Kenya, Tanzania, Mozambique, Malawi and Zimbabwe.

Flowering period August and September.

Conservation status Rare in Ethiopia. Locally common elsewhere in its range.

Notes Differs from *P. aethiopica* by the leaves being less than 6 mm broad.

9. P. caduca *Rchb.f.*

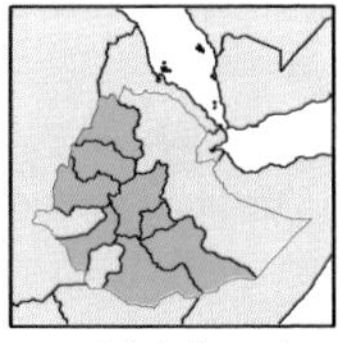

Polystachya caduca

The specific epithet '*caduca*' refers to the caducous bracts. H.G. Reichenbach described it in 1852 from plants collected in Ethiopia by Georg Wilhelm Schimper.

Plant with clustered or sequences of small subglobose to conical or obliquely conical pseudobulbs, 5–20 × 2–10 mm, green drying dull yellow, 2–4-leaved at apex. Leaves linear to narrowly elliptic or oblanceolate, 15–90 × 2–9 mm. Inflorescence erect, laxly 2–8-flowered, 2.5–6 cm long, shorter than the leaves; peduncle slender, 1.5–3 cm long, pubescent; rhachis pubescent. Bracts linear, 2–6 × 1 mm. Flowers white with a yellow mark on the lip and a yellow anther cap; pedicel and ovary 4–5 mm long, pubescent. Dorsal sepal lanceolate, 5–7 × 1.5–2 mm. Lateral sepals very obliquely falcate-ovate, 6–8 × 6–7.5 mm; mentum obliquely conical-cylindrical, 4–5 mm long. Petals oblong-lanceolate, 4–4.5 × 1–1.2 mm, 3-veined. Lip longly clawed, obscurely 3-lobed and cordate-ovate in front, 7–7.5 × 3.5–3.7 mm; midlobe 3.5–3.6 × 3 mm.

Habitat and distribution An epiphyte in montane *Podocarpus* forest and on scattered *Acacia* and *Albizia* in montane grassland between 1820 and 2800 m in Gonder, Gojam, Shewa, Arsi, Bale, Sidamo, Kefa and Wellega. Unknown elsewhere.

P. caduca

Flowering period March to May and August to October.

Conservation status Vulnerable.

Notes Differs from *P. bennettiana* and *P. rivae* in being a much smaller plant, 8 cm or less tall, and with smaller flowers.

10. P. steudneri *Rchb.f.*

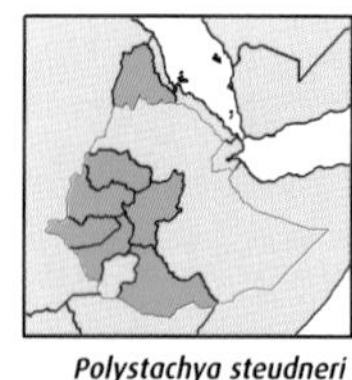

Polystachya steudneri

The specific epithet '*steudneri*' was given in honour of Dr Steudner who collected the type specimen from R. Guang in Ethiopia. H.G. Reichenbach described it in 1881.

Plant 11–35 cm tall. Pseudobulbs narrowly conical, 2–5 × 5–9 mm, 2–3-leaved at apex. Leaves linear to narrowly oblong-lanceolate, up to 12 × 0.5–1.3 cm, apex obscurely unequally bilobed. Inflorescence produced after the leaves have fallen, 8–33 cm long, paniculate; peduncle and rhachis terete; branches short, densely many-flowered, secund. Bracts lanceolate, 1.5–3 × 1 mm. Flowers white, yellow or yellow-green, heavily spotted or lined with crimson. Dorsal sepal lanceolate, 3–6 × 1.6 mm; lateral sepals obliquely ovate-lanceolate, 3.7–4 × 1.8–2.3 mm; mentum up to 2 mm long. Petals oblong-lanceolate, up to 3.5 × 0.7 mm. Lip 3-lobed, shortly clawed, 3 × 2.3 mm; side lobes at

P. steudneri

right-angles to midlobe; midlobe oblong-ovate, 1.6 × 1 mm; fleshy obconical callus on claw of lip.

Habitat and distribution In deciduous woodland or dryish scrub between 1250 and 2300 m in Gojam, Shewa, Sidamo, Illubabor, Kefa and Wellega. Also in Nigeria, Cameroon, Sudan, Uganda and Kenya.

Flowering period April to July and October to January.

Conservation status Vulnerable in Ethiopia because of habitat destruction.

Notes Distinguished from *P. eurychila* by the branched inflorescence.

11. P. eurychila *Summerh.*

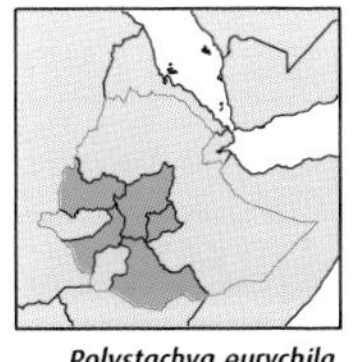

Polystachya eurychila

The epithet '*eurychila*' refers to the broad lip. Victor Summerhayes described it in 1939 from a collection made by Bill Eggeling from Elgon, between Butandiga and Bulambuli, Uganda.

Plant 8–39 cm tall. Stems clustered, somewhat pseudobulbous at base, narrowly conical, up to 20 × 8 mm, 3–5-leaved towards apex. Leaves deciduous,

P. eurychila

suberect, narrowly linear-lanceolate, grass-like, up to 29 × 0.6–0.9 cm. Inflorescence unbranched, mostly borne after the leaves have fallen, 4–11 cm long, subdensely many-flowered; peduncle covered by 2 scarious acute sheaths; rhachis sparsely setulose-pubescent. Bracts broadly deltoid-ovate, 1–2 mm long. Flowers pale lilac-rose with purple to magenta. Dorsal sepal oblong-elliptic, 3.5–4 × 1.7–2 mm; lateral sepals rotund-triangular, dorsally shortly cuspidate below the apex, 4–5 × 4–5 mm; mentum conical, 3–3.5 mm long. Petals ligulate or spathulate-ligulate, 3.5 × 0.6–1.4 mm. Lip shortly clawed, lacking a callus, very broadly 3-lobed in apical half, 3.5–4.5 × 6–7.1 mm, whole surface farinaceous-pubescent; side lobes spreading, elliptic-oblong, 3 × 2–2.5 mm; midlobe broadly triangular, 1–1.5 × 1.5–2 mm.

Habitat and distribution Found in riverine forest or on wet rocks between 1250 and 2700 m in Shewa, Arsi, Sidamo, Kefa and Wellega. Also in Uganda and Kenya.

Flowering period October, December to April.

Conservation status Vulnerable.

Notes Distinguised from *P. steudneri* by its simple unbranched inflorescence.

12. P. cultriformis *(Thouars) Spreng.*

The specific epithet '*cultriformis*' refers to the knife-shaped leaf. The French botanist Louis Marie Aubert du Petit Thouars described it as *Dendrobium cultriforme* in 1822 based on his own collection from Mauritius. Kurt Sprengel transferred it to *Polystachya* in 1826.

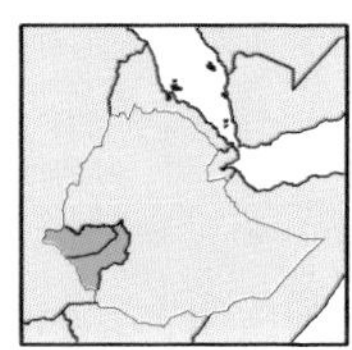

Polystachya cultriformis

Plant 6–24.5 cm tall. Pseudobulbs arising from a short creeping rhizome, loosely clustered, narrowly cylindric to conical, 1.4–18 × 0.2–1.2 cm. Leaf obovate, ovate or elliptic, 3.2–36 × 1.2–5.5 cm, articulated 2–6 mm above the apex of the pseudobulb. Inflorescence usually longer than the leaf, paniculate or rarely racemose, 4.4–29 cm long, bearing up to 50 flowers successively; peduncle slender to stout, 3.2–21.5 cm long. Bracts ovate-triangular, 2.5–4.5 mm long. Flowers white, yellow, green, pink or purple, more or less yellow at the base of the lip. Dorsal sepal ovate-triangular, 4–8 × 2–4.5 mm; lateral sepals obliquely triangular, 5–8 × 3–6 mm; mentum conical, up to 7 mm long. Petals linear to spathulate, 3.5–7.5 × 1–2.5 mm. Lip strongly recurved, more or less shortly clawed, distinctly 3-lobed in the apical half, 4–7.8 × 3–6 mm; side lobes porrect; midlobe oblong-quadrate, 1.2–4.5 × 1.5–3.5 mm; callus fleshy, more or less central.

Habitat and distribution In montane forests, and secondary forest around tea plantations between 1700 and 2200 m in Illubabor and Kefa. Also in Uganda, Kenya, Tanzania, Cameroon, Fernando Po, Gabon, DR Congo, Burundi, Malawi, Mozambique, Zimbabwe, South Africa, Madagascar, Mascarenes and the Seychelles.

Flowering period December and January.

Conservation status Vulnerable in Ethiopia. Locally common elsewhere.

Notes Distinguished from other Ethiopian species in the genus by the solitary leaf at the apex of the pseudobulb.

P. cultriformis

Stolzia grandiflora

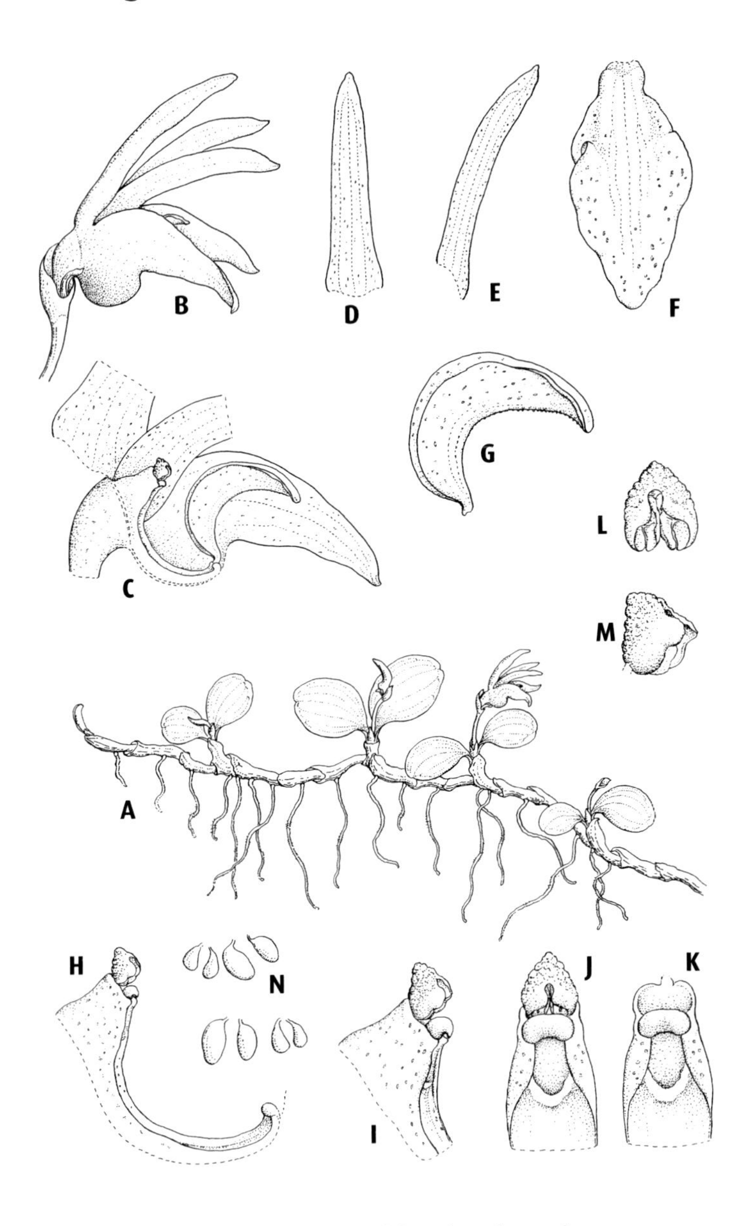

A habit, × 2/3; **B** flower, × 3; **C** flower with lateral sepal & petal removed, × 4; **D** dorsal sepal, × 3; **E** petal, × 3; **F** lip flattened, × 7; **G** lip side view, × 7 ; **H** column, × 8; **I** column apex side view, × 13; **J** column apex front view, × 13. **K** column lacking anther, × 13; **L, M** anther cap, × 16; **N** pollinarium, × 13.
All drawn from *de Wilde* 6093 by Mrs. M. Church.

21. STOLZIA *Schltr.*

Dwarf epiphytic or rarely lithophytic herbs. Stems pseudobulbous, repent, forming mats on the surface of the substrate. Pseudobulbs asymmetrical, fleshy, ovoid, fusiform to clavate, bearing 1–2 leaves at the apex which is often offset. Leaves fleshy or coriaceous, spreading or erect, oval, ligulate or obovate. Inflorescence erect, 1–many-flowered. Flowers more or less secund, somewhat campanulate, yellow, orange, brown, red or green. Lateral sepals fused at base, forming with the column-foot a more or less prominent mentum. Lip entire, curved, V-shaped in cross-section. Column truncate, with a long curved foot more than 3 times as long as the column; pollinia 8, in 2 groups, 4 larger and 4 smaller; stigma concave with a flap-shaped rostellum in front.

A genus of about 15 species, all from tropical Africa. Two species are found in Ethiopia.

Key

1 Lip 4 mm long with a fleshy knob-like callus at the base; leaves 1–2.7 cm long; flowers pale yellow or greenish yellow with a pale brownish-red lip **1. S. grandiflora**

\- Lip 2 mm long, lacking a callus at the base; leaves 0.5–1.4 cm long; flower yellow, brown or reddish, more or less striped red or brown **2. S. repens**

1. S. grandiflora *P.J.Cribb*

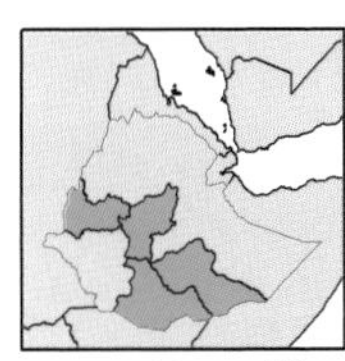

Stolzia grandiflora

The specific epithet '*grandiflora*' refers to the large flowers. Phillip Cribb described it in 1985 from a plant collected by Mike Gilbert at Wonchi Crater Lake in Shoa Province.

A creeping plant; pseudobulbs slender clavate, rhizomatous, rather wiry, swollen below each pair of leaves, 2–4 cm long, up to 3 mm diameter. Leaves elliptic to obovate, 1–2.7 × 0.6–1 cm. Inflorescence terminal, 1-flowered; peduncle 3–8 mm long; bract 2–5 mm long. Flowers pale yellow or greenish yellow with a pale brownish-red lip; ovary 3–4 mm long. Dorsal sepal 9–10.5 × 2.5–3 mm; lateral sepals 8 × 3.5 mm; forming with the column foot a short, incurved-conical mentum 2–4 mm long. Petals 9 × 1.5–2 mm. Lip with an obscure fleshy knob-like callus at base, 4 × 2 mm.

S. repen

Habitat and distribution An epiphyte at the edge of montane forest and swamp forest between 1900 and 2850 m in Shewa, Bale, Sidamo and Wellega. Unknown elsewhere.

Flowering period November to February, May and April.

Conservation status Vulnerable because of habitat destruction.

Notes *S. grandiflora* differs from *S. repens* in having larger flowers, with sepals at least 9 mm long and the lip 4 mm long.

2. S. repens *(Rolfe) Summerh.*

The specific epithet '*repens*' refers to the creeping habit of the plant. Robert Rolfe described it as *Polystachya repens* in 1912 from a plant collected in Uganda by E. Brown. Victor Summerhayes transferred it to the present genus in 1953.

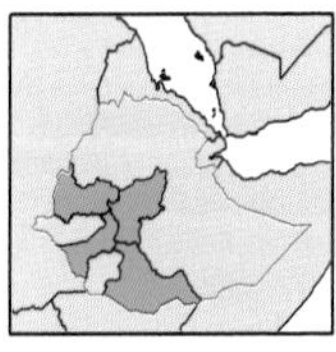

Stolzia repens

Plant to 1 cm high. Pseudobulbs prostrate except at apex, elongate-clavate or rhizomatous, to 3 × 0.3 cm, bearing 2 leaves near insertion of next pseudobulb. Leaves 2, elliptic or obovate, 0.5–1.4 × 0.3–0.8 cm, rounded. Inflorescence 1-flowered, borne between leaves on very short peduncle. Flower yellow, brown or reddish, more or less striped red or brown. Dorsal sepal to 7 × 2.5 mm; lateral sepals to 6 × 2.4 mm, united towards base with each other and the column-foot to form a saccate mentum. Petals to 5 × 1.5 mm. Lip 2 × 1 mm, slightly papillate below.

Habitat and distribution On mossy trunks and branches in forest between 900 and 2750 m in Shewa, Sidamo, Kefa and Wellega. Also in Ghana, Nigeria, Cameroon, DR Congo, Uganda, Kenya, Tanzania, Zambia, Malawi and Zimbabwe.

Flowering period January to March.

Conservation status Rare but possibly overlooked in Ethiopia. Locally common elsewhere in its range.

Notes Differs from *S. grandiflora* in its much smaller orange-red flowers with sepals up to 7 mm long and the lip 2 mm long.

22. BULBOPHYLLUM *Thouars*

Epiphytic or lithophytic sympodial herbs; rhizome short or long and creeping, often branched, slender to quite stout. Pseudobulbs 1-noded, clustered or distant, stem-like or more often swollen, more or less angled in cross-section, arising from rhizome at intervals, 1–2-leaved (rarely more) at the apex. Leaves mostly coriaceous or fleshy or rarely thin-textured, small to large. Inflorescences arising from base of pseudobulb or rarely from node on rhizome, 1–many-flowered, racemose or rarely umbellate. Flowers more or less fleshy, often not opening widely, mostly small in African species, white, cream or green to orange and purple. Dorsal sepal subequal or less commonly much shorter than lateral sepals; lateral sepals free or rarely adnate, united at base to column-foot to form a more or less prominent mentum. Petals mostly much smaller than sepals. Lip often much smaller than sepals, hinged to end of column-foot, often highly mobile, often fleshy, ligulate and curved, mostly entire. Column short, often winged and with apical stelidia; anther small, 2-chambered, with 4 pollinia in pairs; column-foot mostly incurved, united to base of lateral sepals.

A large genus of perhaps 1500 species with a circumtropical distribution. About 100 species are found in tropical Africa and a further 160 or more in Madagascar. It is represented by four species in Ethiopia.

Key

1	Floral bracts 1.5 times as long as flower (including pedicel and ovary)	**3. B. lupulinum**
–	Floral bracts much shorter than flower (including pedicel and ovary)	2
2	Pseudobulbs with 2 leaves	**4. B. scaberulum**
–	Pseudobulbs with 1 leaf	3
3	Inflorescence more than 4 times as long as leaf; leaf 1–5 cm long; lip dentate-ciliate	**1. B. intertextum**
–	Inflorescence less than twice the length of the leaf; leaf 8–16 cm long; lip papillose but not dentate-ciliate	**2. B. josephi**

B. josephi

1. B. intertextum *Lindl.*

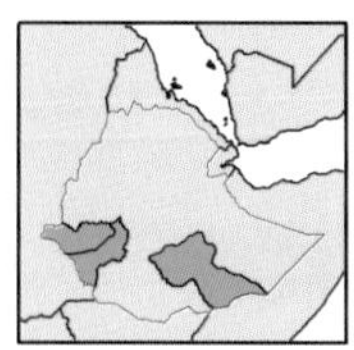
Bulbophyllum intertextum

The specific epithet '*intertextum*' refers to the interwoven rhizomes. John Lindley described it in 1862 from a plant collected in Nigeria by Gustav Mann.

Plant up to 5 cm tall. Pseudobulbs 1-leafed, clustered on very short rhizome, ovoid, 3–10 × 3–7 mm. Leaf linear-oblong, obtuse, rounded or subacute, 1–5 × 3–7 cm. Inflorescence 5–10 cm long; peduncle and rachis wiry. Bracts 2–2.5 mm long. Flowers greenish cream to purple-red. Dorsal sepal 3.5 × 2 mm; lateral sepals 4 × 1.8 mm. Petals 2 × 1 mm. Lip obscurely 3-lobed in basal half, 1.8 × 1 mm, 2-ridged in basal half; side lobes auricular, dentate-ciliate; midlobe minutely ciliate, papillate.

Habitat and distribution Found as an epiphyte in rainforest, riverine and moist forest between 300 and 1800 m in Bale, Illubabor and Kefa. Also in Sierra Leone to Cameroon, Kenya, Tanzania, Angola to Malawi and Zimbabwe and the Seychelles.

Flowering period October to December.

Conservation status Vulnerable in Ethiopia but widespread elsewhere.

Notes Differs from *B. josephi* in being a much smaller plant with an erect inflorescence over 4 times as long as the leaf and a dentate-ciliate lip.

2. B. josephi *(Kuntze) Summerh.*

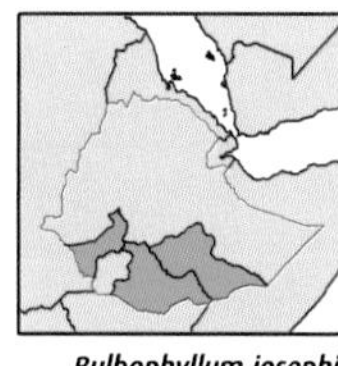
Bulbophyllum josephi

The specific epithet '*josephi*' was given in honour of Sir Joseph Hooker. Otto Kuntze described it as *Phyllorchis josephi* in 1891 from a plant collected in West Africa by Gustav Mann. Victor Summerhayes transferred it to *Bulbophyllum* in 1945.

Plant with a short stout creeping rhizome. Pseudobulbs clustered, 1-leafed, conical-ovoid, wrinkled when dry, 1.4–3.6 × 0.7–2.3 cm, reddish, green or brownish green, turning orange-yellow on drying. Leaf narrowly oblong or oblanceolate, 8–16 × 1–2.4 cm. Inflorescence pendent, 8–26 cm long, densely many-flowered. Bracts 5–6 mm long. Flowers white, greenish white or yellow more or less tipped with crimson or pink. Dorsal sepal 7–7.5 × 1 mm; lateral sepals 7–8 × 1.5–2 mm. Petals 3 × 1.2 mm, papillose in apical half. Lip recurved, 2.2 × 1.3 mm, with 2 longitudinal central ridges, margins papillose.

B. josephi

Habitat and distribution An epiphyte in montane forest between 850 and 2100 m in Bale, Sidamo and Kefa. Also in DR Congo, Rwanda, Uganda, Kenya, Tanzania, Zambia, Malawi, Mozambique and Zimbabwe.

Flowering period September to December.

Conservation status Vulnerable in Ethiopia but locally common elsewhere in its range.

Notes Differs from *B. intertextum* by the longer, pendent, more densely flowered inflorescence which is twice as long as the leaf and the papillose lip.

3. B. lupulinum *Lindl.*

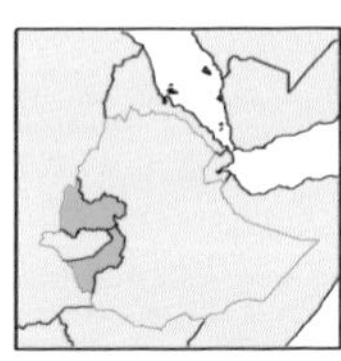

Bulbophyllum lupulinum

The specific epithet '*lupulinum*' refers to the inflorescence, in which the bracts resemble those of the hop (*Humulus lupulus*). John Lindley described it in 1862 from a plant collected in West Africa by Gustav Mann.

Pseudobulbs 1–2-leaved, ovoid to narrowly ovoid or ellipsoid, 2.7–7.5 × 1.2–2.5 cm, slightly flattened, sharply 4-angled. Petiole 2–15 mm long. Leaf lanceolate to linear-lanceolate, 8–23 × 1.2–5 cm. Inflorescence 15–38 cm long, 28–68-flowered. Peduncle 65–190 × 2.5–5 mm. Rhachis 5–23 cm long; surface with dots of fine dark hairs. Floral bracts 9.5–16 × 7–13 mm. Flowers distichous, purplish. Pedicel and ovary 1.3–2.2 mm long, with fine dark hairs. Dorsal sepal broadly to narrowly triangular, 4–5.2 × 1.5–3 mm. Lateral sepals free, 3.8–5.2 × 1.5–2.9 mm. Petals 2.5–3.2 × 0.25–0.5 mm; margins entire. Lip with a reflexed top part, more or less rectangular in outline (not spread), 1.5–2.4 × 0.8–1.7 mm.

Habitat and distribution An epiphyte in lowland rainforest to montane forest, semideciduous forest, also frequently found as a lithophyte, even in exposed places such as rocky mountain slopes between 1000 and 1700 m in Kefa and Wellega. Also in Guinea Bissau, Nigeria, Cameroon, DR Congo and Zambia.

Flowering period November.

Conservation status Rare and endangered in Ethiopia, where it is known only from a single collection. Locally common elsewhere in its range.

Notes *B. lupulinum* differs from *B. josephi, B. intertextum* and *B. scaberulum* by its distinctive inflorescence in which the floral bracts are 1.5 times as long as the flower.

B. lupulinum

4. B. scaberulum *(Rolfe) Bolus*

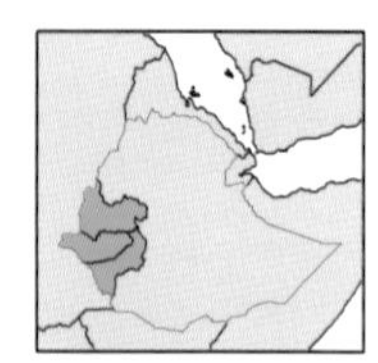

Bulbophyllum scaberulum

The specific epithet '*scaberulum*' refers to the rough surface of the sepals and ovary. Robert Rolfe described it as *Megaclinium scaberulum* in 1888 from a plant collected by Tillett in Pondoland, South Africa. Harry Bolus transferred it to *Bulbophyllum* in 1889.

Plant with a stout creeping rhizome 2–3 mm diameter. Pseudobulbs 2-leaved, ovoid, narrowly ovoid or ellipsoid, 4–5-angled, 2–5.5 × 0.8–1.3 cm, at 2–10 cm intervals on rhizome. Leaves 4.5–12.5 × 1–1.9 cm. Inflorescence erect, 5.5–30 cm long, many-flowered; rhachis bilaterally flattened, more or less curved like a bow, 3–8 mm broad, with a scabrid indumentum. Bracts 3–4 mm long. Flowers scabrid, maroon, green or ochre, heavily marked with maroon. Dorsal sepal 6–7 × 1.5 mm; lateral sepals 4–5.5 × 2–3 mm. Petals 3–4 × 0.3–0.4 mm. Lip deflexed, 2–2.5 × 1.8–2 mm.

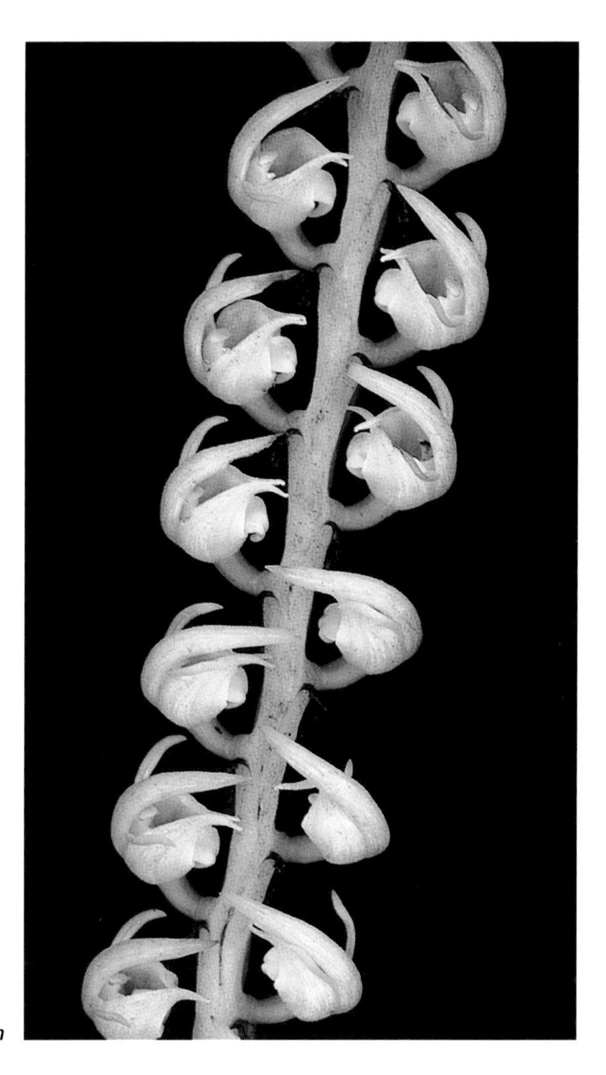

B. scaberulum

Habitat and distribution An epiphyte in riverine and montane forests between 1000 and 1800 m in Kefa, Illubabor and Wellega. Also widespread in tropical and southern Africa.

Flowering period June; November and December.

Conservation status Vulnerable in Ethiopia but locally common elsewhere in its range.

Notes Differs from all other *Bulbophyllum* species in Ethiopia by its bilaterally flattened rhachis on which the flowers are borne on the mid-vein of each side.

23. OECEOCLADES *Lindl.*

Terrestrial, rarely epiphytic herbs. Pseudobulbs close together, usually ovoid to fusiform, more or less approximate, usually heteroblastic (with only one internode elongated, the remaining basal ones very short), apex 1–3-leaved, up to 15 × 3 cm, but often narrower. Leaves usually with duplicate venation, coriaceous, conduplicate, often variegated, usually petiolate, the petiole articulate some distance above the base and sometimes above the middle, the line of articulation consisting of a number of irregular blunt or acute teeth or occasionally more or less regular. Inflorescences arising from the base of the pseudobulb, often exceeding the leaves, simply racemose or frequently paniculate; bracts inconspicuous, rarely with a basal extrafloral nectary. Flowers resupinate, rather small, thin in texture. Sepals and petals free, variously spreading, similar, the petals usually slightly shorter and broader. Lip decurved, spurred, 3- or apparently 4-lobed; side lobes erect; midlobe usually lobulate or emarginate; disc either with 2 approximate, quadrate or triangular calli at the spur entrance or with 3 variously thickened, parallel ridges which together with the lateral nerves are sparsely but distinctly hirsute. Column erect, short, oblique at the base or with a short foot; anther cucullate or cristate; pollinia 2, ovoid or pyriform, on a short or rudimentary stipe; viscidium large; stigmata confluent; rostellum short.

A genus of some 31 species in tropical and South Africa, Madagascar and the Mascarene Islands with a single widespread species in tropical America. It is represented by two species in Ethiopia.

Key

1 Plant 40–80 cm tall; pseudobulbs caespitose; leaves 10–26 × 3–8 cm, petiole 6–7 cm long; inflorescence with extrafloral nectaries exuding viscid liquid; flowers greenish yellow **1. O. saundersiana**

– Plant 12–26 cm tall; pseudobulbs superposed; leaves 6–11 × 2.5–4.5 cm; petiole 2–6 cm long; inflorescence without extrafloral nectaries, flowers mostly white with purple on lip **2. O. ugandae**

O. saundersiana

1. O. saundersiana *(Rchb.f.) Garay & P.Taylor*

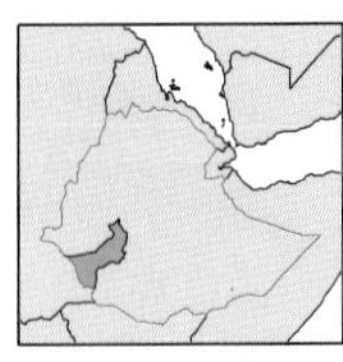
Oeceoclades saundersiana

The specific epithet was given in honour of the well-known Victorian orchid grower William Saunders who flowered the type plant. H.G. Reichenbach described it as *Eulophia saundersiana* in 1866 from a plant collected in West Africa by Gustav Mann. Leslie Garay and Peter Taylor transferred it to *Oeceoclades* in 1976.

Plant 40–80 cm tall, glabrous. Pseudobulbs caespitose, narrowly ovoid to fusiform, very slightly compressed, obscurely sulcate, erect or spreading, 4–15 cm long, the widest part up to 2.5 cm broad, dark, slightly glossy green, 2- or occasionally 1-leaved. Leaves narrowly elliptic or ovate-elliptic, sometimes oblique; lamina 10–26 × 3–8 cm; petiole irregularly articulated, 6–17cm long. Inflorescence simple or with 1 or 2 short branches at the base, 16–30 × 4–5 cm, dense or lax, 15–28-flowered. Bracts 2–8 mm long, with a basal extrafloral nectary exuding copious amounts of viscid sweet liquid. Flowers porrect at first, becoming somewhat deflexed with age; sepals and petals pale greenish yellow; lip pale yellow or greenish cream, green towards the spur-entrance; calli creamy white; spur reddish brown; column green, anther yellow; pedicel and ovary slender, 12–18 mm long, reddish brown. Dorsal sepal 8.5–14 × 3–5 mm. Lateral sepals 9.7–14 × 3–5 mm. Petals 8–13 × 4.5–7 mm. Lip bent forwards from and articulated at the base, distinctly 3-lobed, mostly glabrous, 10–15 mm long; midlobe divided into 2 lobules, each oblong to oblong-orbicular, usually divaricate, longer than side lobes, 8–12 × 7–14 mm; side lobes more or less orbicular; callus composed of 2 more or less parallel fleshy, glabrous keels, shorty pubescent at the spur-entrance and at the apex; spur cylindrical, 4–6 mm long.

Habitat and distribution In damp shady places in forest and thickets between 1200 and 1300 m in the Kefa region in Ethiopia. Also in Sierra Leone, Liberia, Ivory Coast, Ghana, Nigeria, Cameroon, Gabon, Uganda, Kenya, Tanzania, DR Congo, Zambia and Angola.

Flowering period January and February.

Conservation status Rare and vulnerable.

Notes Distinguished from *O. ugandae* by its simple or sparsely branched inflorescence and larger greenish-yellow flowers.

2. O. ugandae *(Rolfe) Garay & P.Taylor*

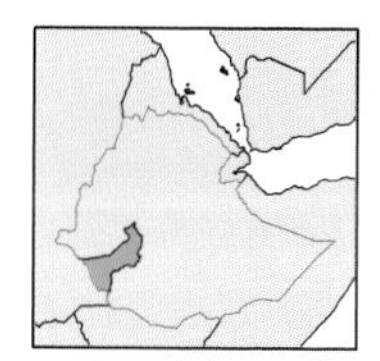

Oeceoclades ugandae

The specific epithet '*ugandae*' refers to Uganda from where the type specimen was collected. Robert Rolfe described it as *Eulophia ugandae* in 1913 from a plant collected by E. Brown in Mengo District, Uganda. Leslie Garay and Peter Taylor transferred it to the genus *Oeceoclades* in 1976.

Plant 12–26 cm tall. Stem climbing. Pseudobulbs superposed, more or less fusiform, 75–110 × 5–8 mm, 2–3-leaved. Leaves narrowly elliptic or ovate-elliptic; lamina 6–11 × 2.5–4.5 cm; petiole irregularly articulated, 2–6 cm long. Inflorescence paniculate, compact, 6–14 × 2–3 cm, many-flowered, lax or subdense. Bracts 4–15 mm long. Flowers small; sepals and petals white, sometimes flushed green; lip white, with a few purple nerves and a purple throat; pedicel and ovary 6–13 mm long. Dorsal sepal 6.5–8.5 × 2 mm; lateral sepals 7–9 × 2–2.5 mm. Petals 6–7.8 × 1.5–2.3 mm. Lip distinctly 3-lobed, with very short hairs at the spur entrance and along the nerves, otherwise glabrous, 7–8 × 4.3–6.2 mm; midlobe 2.7–3 × 4 mm; side lobes oblong; spur short, clavate, much dilated and globose at the apex, 3–4 mm long.

Habitat and distribution A terrestrial, occasionally epiphytic herb in primary forest between 1000 and 1200 m in Kefa. Also in Ghana, DR Congo and Uganda.

Flowering period September to December.

Conservation status Rare and threatened.

Notes Readily identified by the paniculate inflorescence and small flowers.

O. ugandae

24. PTEROGLOSSASPIS *Rchb.f.*

Terrestrial, sympodial herbs. Subterranean stem composed of a series of lobed fleshy tubers; aerial stem very short, leaves 1–3, linear-lanceolate, long-petiolate, plicate, enclosed in sheaths towards base, arising with scape from the current year's growth. Scape erect, covered with subscarious, tubular sheaths. Inflorescence a several-flowered, short dense, rarely elongated and lax, terminal raceme; bracts usually conspicuous, long, narrow, rarely broad, scarious, often exceeding the flowers. Flowers resupinate or non-resupinate, medium-sized. Sepals and petals similar, free, usually spreading, the petals usually shorter than the sepals. Lip sessile, 3-lobed, rarely entire, neither spurred, saccate nor clawed, flat, never concave, midlobe larger than side lobes, apex usually recurved, disc either entirely tuberculate or the nerves variously tuberculate, verruculose or lacerately keeled, median nerve of the side lobes often raised and ridge-like at the base. Column very short, rarely elongated, broad, curved and biauriculate at the base, the 2 auricles merging with the base of the lip, wings absent, foot absent or very rarely rudimentary, more or less horizontal; anther terminal, ovate, obtuse, apiculate, unilocular; pollinia 2, subglobose, united on a short broad stipe; viscidium large, scale-like.

A genus of seven species, five in tropical and subtropical Africa, with one in south-eastern USA and Cuba and another in Argentina and Brazil. A single species is reported from Ethiopia.

P. eustachya *Rchb.f.*

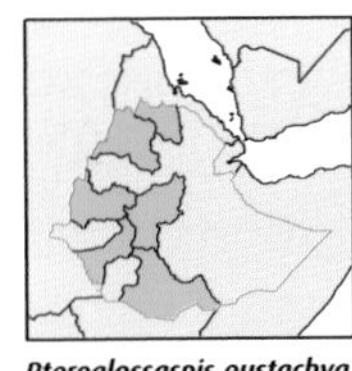
Pteroglossaspis eustachya

The specific epithet '*eustachya*' refers to the true stem. H.G. Reichenbach described it in 1878 from a plant collected by Georg Wilhelm Schimper in Gonder/Tigray.

Plant 22–65 cm tall, glabrous. Leaves linear to linear-lanceolate; lamina 26–40 × 1.5–2 cm; petiole 12–18 cm long. Inflorescence cylindrical or ovoid, many-flowered, dense; rhachis 2–8 cm long. Bracts pale brown, lowermost 3–8 cm long, uppermost shorter. Flowers resupinate, variable in colour; sepals and petals white, dull yellow, suffused maroon or entirely purple-maroon; midlobe of lip dark purple-maroon or crimson, side lobes yellow or pale purple-red with darker purple-red veins; pedicel and ovary 1–2 cm long. Dorsal sepal 10–11 × 3.5–4 mm. Lateral sepals

P. eustachya

10 × 4–4.5 mm. Petals 9–11 × 3.5–4 mm. Lip 3-lobed, 7–9 × 12–14 mm; midlobe 6–8.5 mm long, divaricate, 5–6 mm long.

Habitat and distribution In wooded grassland, damp grassland, stream-sides, short-grassed lava plains between 1500 and 2000 m in Tigray, Gonder, Shewa, Sidamo, Kefa and Wellega. Also in Uganda, Kenya, Tanzania, Mozambique and Zimbabwe.

Flowering period July to October.

Conservation status Vulnerable.

Notes Could be confused with some *Eulophia* species but the lip lacks a spur and is 3-lobed in the basal half.

25. EULOPHIA *Lindl.*

Small, medium or large terrestrial or rarely lithophytic herbs; roots slender to stout, basal or adventitious, often with a well-defined velamen. Perennating organs stem-like, pseudobulbous or tuber-like, above the ground or more commonly underground, conical, cylindrical or irregular in shape, several-noded. Leaves usually present and green but in some species much reduced, scale-like and brown or buff; green leaves 1–many, thin-textured to fleshy or coriaceous, with or without prominent longitudinal veins, linear, lanceolate, ovate or elliptic, sheathing at the base; scale leaves sheathing when present. Inflorescences basal, laxly to subdensely many-flowered, usually racemose, rarely branching. Flowers small to large, sometimes showy and brightly coloured. Sepals and petals similar or with the petals much broader, free to base or with the lateral sepals fused at the base to the column-foot. Lip 3-lobed, usually spurred at the base, usually with a callus of ridges and/or papillae on upper surface. Column short to long, with or without a column-foot; anther-cap entire or 2-lobed at the apex; pollinia 2, subglobose; stipes solitary, triangular to oblong; viscidium oblong, elliptic or lunate.

A genus of about 250 species, widespread in tropical and southern Africa, Madagascar, the Mascarenes, tropical and subtropical Asia, SE Asia, Australasia and the tropical Americas. Twenty-five species are reported from Ethiopia.

E. cucullata

Key

1	Lip entire except for small basal auricles, ovate, lacking a callus; spur at least 15 mm long	**2. E. guineensis**
–	Lip three-lobed, callose; spur obscure up to 7 mm long	2
2	Pseudobulbs conspicuous, above ground, leafy, covered by leaf-bases or naked	3
–	Pseudobulbs underground, tuberous	5
3	Pseudobulbs not covered by leaf sheaths, green turning yellow or orange with age, 2–3-leaved; leaves fleshy-coriaceous, tapering from base to apex, serrate; inflorescence often branching; sepals and petals recurved strongly towards apex	**3. E. petersii**
–	Pseudobulbs covered by leaf bases in first season; leaves plicate, linear or lanceolate, not serrate; inflorescence unbranched; sepals and petals not recurved in apical part	4
4	Pseudobulbs ovoid; sepals and petals markedly different in shape and colour; sepals oblong-obovate, obtuse; petals elliptic to semicircular, obtuse, yellow; lip yellow with a convex reduplicate midlobe; spur 2–3.5 mm long	**4. E. streptopetala**
–	Pseudobulbs cylindrical; sepals and petals subsimilar, lanceolate, acuminate; petals lanceolate, green with a brownish apex or purplish brown; lip midlobe flat to slightly concave, with or without undulate margins, white with a pink or purple band across the base; spur 5–7 mm long, somewhat clavate	**1. E. euglossa**
5	Lip virtually lacking a spur or saccate base; column-foot obscure	**20. E. albobrunnea**
–	Lip with a conical to cylindrical spur or with a saccate base; column-foot obvious	6
6	Base of lip distinctly saccate; callus on lip with two quadrate erect lamellae in centre	7
–	Base of lip more or less spurred; calli on lip not quadrate	8
7	Lip 18–34 × 20–40 mm; petals elliptic or obovate, 15–24 mm long, 10–21 mm wide	**12. E. cucullata**
–	Lip 14–19 × 8–12 mm; petals oblong-ligulate, 15–19 mm long, 3–5 mm wide	**5. E. alta**
8	Petals elliptic to subcircular, less than twice as long as broad	
–	Petals oblong, ovate or lanceolate, more than twice as long as broad	17

E. guineensis

9	Sepals and petals 5–7.5 mm long	**18. E. pyrophila**
–	Sepals and petals at least 8 mm long	
10	Lip midlobe flat or concave; petals 19–35 × 10–18 mm; lip 20–40 mm long	15
–	Lip midlobe reduplicate, convex; petals and lip not as above	11
11	Flowers with a yellow lip, sometimes marked with red; spur often obscure	12
–	Flowers with white or rose-coloured sepals and petals and a purple or pink lip; spur obvious	14
12	Flowers yellow, sometimes with red venation on side lobes of lip; spur 1–3 mm long	**15. E. speciosa**
–	Flowers yellow and red, the petals strongly marked with red on inner side; spur 8–16 mm long	13
13	Spur 13–16 mm long, pendent, narrowly conical	**16. E. orthoplectra**
–	Spur short, slightly upcurved, 8–9 mm long	**17. E. schweinfurthii**
14	Lip midlobe with cristate-papillose raised ridges	**13. E. cristata**
–	Lip with smooth ridges	**14. E. livingstoniana**
15	Sepals obovate or obliquely obovate; petals and lip purple; callus in centre of lip, comprising 3 raised lamellae	**8. E. horsfallii**
–	Sepals spathulate; petals and lip not as above	16
16	Petals and lip yellow; callus of 3 tall crenulate ridges in basal part of lip, flexuose in front	**9. E. angolensis**
–	Petals and lip white to pink-purple; callus of 3 basal verrucose ridges merging into 5 cristate-papillose crests above	**10. E. caricifolia**
17	Lip lacking verrucae, papillae or hairs on midlobe; sepals, petals and lip pale yellow	**19. E. abyssinica**
–	Lip verrucose, papillate or hairy on midlobe	18
18	Sepals and petals yellow; lip yellow with red papillae; peduncle covered by papery sheaths	19
–	Sepals and petals green, sometimes mottled with brown or reddish; lip white or greenish, often marked with purple; peduncle bearing well-spaced green sheaths	21
19	Sepals and petals at least 20 mm long, lemon-yellow; lip with a maroon mark in the throat	**23. E. zeyheri**
–	Sepals and petals up to 14 mm long, buttercup to canary yellow; lip lacking a maroon mark in the throat	20
20	Sepals and petals 9–14 mm long; lip marked with red papillae; flowering stems covered by papery sheaths	**22. E. odontoglossa**

–	Sepals and petals 5–8 mm long; lip yellow sometimes with reddish tinged side lobes; flowering stems bearing well-spaced sheaths	**21. E. milnei**
21	Sepals and petals 16–34 mm long	22
–	Sepals and petals 8–15 mm long	24
22	Callus of 7–11 verrucose ridges on flabellate midlobe	**11. E. flavopurpurea**
–	Callus papillate on ovate to oblong midlobe	23
23	Flowers spreading; lip 14–15 mm long; callus of 3 ridges, purple-papillate in front	**6. E. stachyodes**
–	Flowers subnutant; lip 20–21 mm long; callus of 2 basal ridges, greenish papillate in front	**7. E. adenoglossa**
24	Lip 9–15 mm long; callus of 3 fleshy ridges, papillate in front	**25. E. kyimbilae**
–	Lip 7.5–10 mm long; callus of 2 ridges, papillate in front	**24. E. clavicornis** var. **nutans**

1. E. euglossa *(Rchb.f.) Rchb.f.*

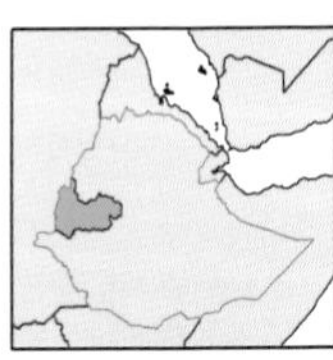
Eulophia euglossa

The specific epithet '*euglossa*' refers to the tongue-like lip. H.G. Reichenbach described it as *Galeandra euglossa* in 1852 from the type speciemen collected in Sierra Leone and cultivated by Oswald van Heer. He transferred it to *Eulophia* in 1866.

Plant 60–150 cm tall. Perennating organs above ground, pseudobulbous, cylindrical-conical, swollen at base, 5–10-noded, 16–25 × 1–1.5 cm. Leaves 5–10, plicate, ovate to lanceolate, 16–42 × 2.5–5.5 cm. Inflorescence laxly many-flowered; peduncle stout, 10–12 mm diameter; rhachis 10–38 cm long. Bracts spreading, lanceolate, 15–36 mm long. Flowers not opening widely; sepals green with a brownish apex; petals similar; lip white with a pink or purple band across base of midlobe and a greenish spur; pedicel and ovary 17–21 mm long. Dorsal sepal lanceolate, 20–24 × 3 mm; lateral sepals similar, 18–24 × 3 mm. Petals lanceolate, 15–18 × 3 mm. Lip 3-lobed, 13–14 × 7.5–9 mm; side lobes narrow; midlobe elliptic, 8 × 6.5 mm, margins undulate; callus of 2–3 obscure ridges in basal half of midlobe; spur clavate, 5–7 mm long.

Habitat and distribution In dense forest on red-brown sandy loam over ironstone; also in secondary forest between 1200 and 1300 m in Wellega. Also in Uganda, Sierra Leone, Liberia, Ivory Coast, Nigeria, Cameroon and Gabon.

Flowering period Not known for Ethiopia.

Conservation status Vulnerable.

Notes Listed as occurring in Ethiopia by Summerhayes (1968) but only one Ethiopian specimen was seen during the preparation of this account. *Eulophia euglossa* differs from *E. guineensis* by its green and white flowers and the 5–7 mm long spur being much shorter.

2. E. guineensis *Lindl.*

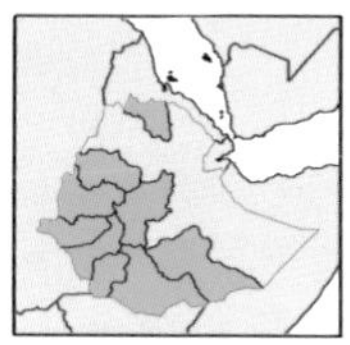

Eulophia guineensis

The specific epithet '*guineensis*' refers to the general area "Guinea" which includes Sierra Leone, from where the type specimen was collected. John Lindley described it in 1823 from a plant collected by George Don.

A terrestrial or lithophytic herb 30–65 cm tall. Perennating organs above ground, pseudobulbous, conical, proximate, 3–3.5 × 1.5–2 cm. Leaves 3–4, plicate, elliptic, 10–35 × 3–9.5 cm, shortly petiolate, articulated to conduplicate leaf-base. Inflorescence laxly 5–many-flowered, opening before or with leaves; peduncle 4–9 mm diameter; rhachis 7–29 cm long. Bracts lanceolate, 10–30 × 2–4 mm. Flowers subnutant, showy; sepals and petals purplish brown; lip pinkish purple with a paler or white base and spur; pedicel and ovary 13–25 mm long. Dorsal sepal linear-lanceolate, 16–26 × 3–4.5 mm; lateral sepals similar. Petals linear-lanceolate, 15–20 × 4–5 mm. Lip obscurely 3-lobed at base, obovate, 20–35 × 13–32 mm; midlobe 20–26 × 17–32 mm; callus absent; spur very slender at apex, 15–25 mm long.

E. guineensis

Eulophia lips (flattened) and spurs

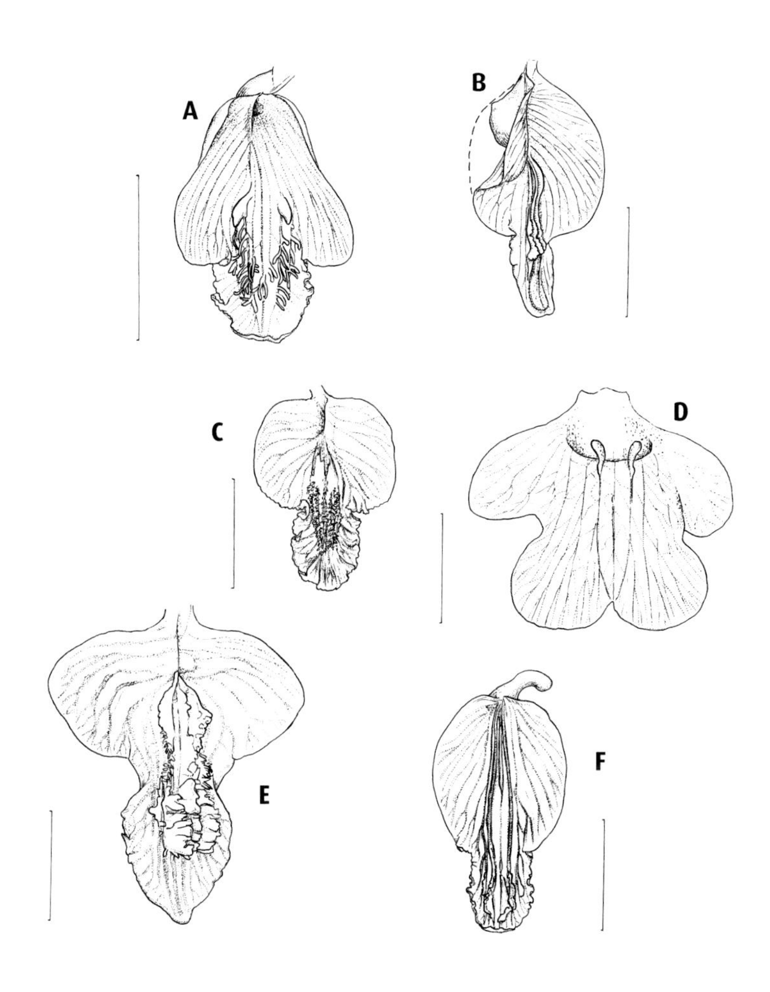

A *E. alta;* **B** *E. angolensis;* **C** *E. caricifolia;* **D** *E. cucullata;* **E** *E. horsfallii;*
F *E. petersii.* Scale bar = 10 mm.

Habitat and distribution In shade or semi-shade amongst rocks, in open scrub, woodland, bushy meadow, and bamboo thicket between 550 and 2000 m Tigray, Gojam, Shewa, Bale, Sidamo, Gamo Gofa, Illubabor, Kefa and Wellega. Also inYemen and Oman, tropical Africa from Sierra Leone across to Uganda, Kenya and Tanzania, and south to Malawi and Zambia.

Flowering period March to August.

Conservation status Vulnerable.

Notes A very distinctive orchid, readily identifiable by the large, ovate, purple lip and long narrow spur.

3. E. petersii *Rchb.f.*

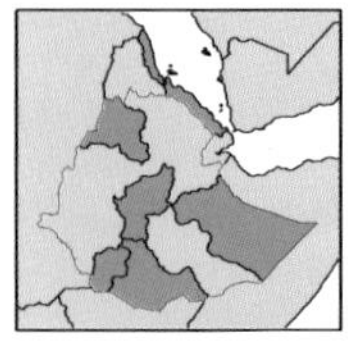

Eulophia petersii

The specific epithet was given in honour of Carl Peters who collected the type specimen from Mozambique. H.G. Reichenbach described it as *Galeandra petersii* in 1847. He transferred it to the present genus in 1865.

A large terrestrial herb 100–350 cm tall. Perennating organs pseudobulbous, 6–23 × 1.4–4 cm, 4–6-noded. Leaves 2–4, somewhat conduplicate, linear-ligulate,

E. petersii

14–80 × 1.4–6 cm, serrate on margins. Inflorescence 3–7-branched, laxly many-flowered; branches to 35 cm long; peduncle stout, up to 90 × 1–1.4 cm. Bracts lanceolate, 5–16 × 4–5 mm. Flowers fleshy; sepals olive-green flushed or striped maroon; lip white, veined red or purple; pedicel and ovary 20–26 mm long. Dorsal sepal erect, oblanceolate, 17–23 × 4–5 mm; lateral sepals similar, 18–24 × 5–7 mm. Petals oblanceolate or oblong, 16–18 mm long. Lip 3-lobed, 14–20 × 8–15 mm; side lobes elliptic; midlobe circular to subquadrate, 4 × 6 mm; callus of 3 fleshy ridges, raised and erose on the midlobe; spur cylindric-subclavate, 4–6 mm long.

Habitat and distribution By sea in sandy soil, in rocky places and on volcanic rocks in *Acacia-Commiphora*, *Grewia* and other thickets and bushland between 1200 and 1800 m in Gonder, Shewa, Harerge, Sidamo and Gamo Gofa and in Eritrea. Also in Arabia, Uganda, Kenya, Tanzania, Zanzibar, DR congo, Sudan south to South Africa.

Flowering period January to April.

Conservation status Vulnerable.

Notes Readily recognised by the bare pseudobulbs topped by 2 or 3 fleshy leaves with saw-toothed edges and by the branching inflorescence up to 3.5 m tall.

4. E. streptopetala *Lindl.*

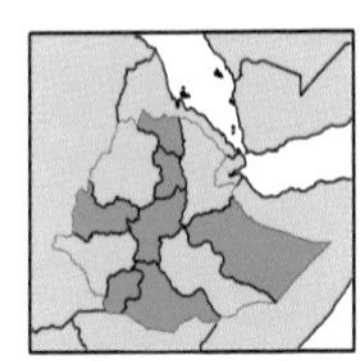

Eulophia streptopetala

The specific epithet '*streptopetala*' refers to the curved petals. John Lindley described it in 1826 from a South African plant cultivated by William Colvill.

Plant 50–150 cm tall. Perennating organs pseudobulbous, conical to cylindric-fusiform, 2.5–10 × 1.7–2.5 cm. Leaves 4–9, plicate, in a fan, lanceolate, 40–65 × 0.4–8 cm. Inflorescence lax; peduncle 9 mm diameter; rhachis 15–35 cm long. Bracts lanceolate or ovate, 7–20 × 5–8 mm. Flowers spreading; sepals green or dull yellow, veined and mottled with brown; petals yellow; lip darker yellow with reddish purple side lobes and a pale purple spur; column white; pedicel and ovary 15–22 mm long. Dorsal sepal oblanceolate to elliptic, 8–18 × 3–9.5 mm; lateral sepals similar but longer. Petals elliptic, circular or obovate, 8–21 × 6–17 mm. Lip difficult to flatten, 3-lobed, 8–12 ×

E. streptopetala

7.5–11 mm; side lobes erect, rounded; midlobe conduplicate, obovate or elliptic, 7.5–12 mm long; callus of 3–5 low fleshy rugulose ridges on midlobe; spur tapering to apex, 2–3.5 mm long.

Habitat and distribution On rocky limestone slopes with scattered shrubs, mixed shrubby slopes, along creeks on mountain slopes with juniper trees, and with *Eucalyptus* between 1300 and 2900 m in Welo, Shewa, Harerge, Sidamo, Gamo Gofa and Wellega. Also in SW Arabia and throughout tropical and southern Africa.

Flowering period March to May.

Conservation status Vulnerable.

Notes Readily recognised by the broadly plicate leaves and flowers with obovate brown chequered sepals, almost circular yellow petals and yellow lip

5. E. alta *(L.) Fawc. & Rendle*

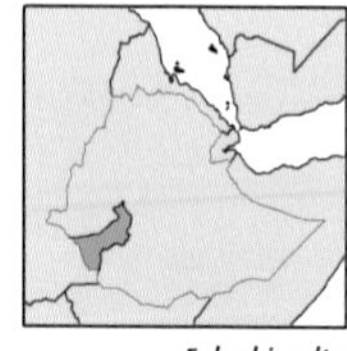

Eulophia alta

The specific epithet '*alta*' refers to the height of the plant. Linnaeus described it as *Limodorum altum* in 1767, based on Charles Plumier's *Limodorum foliis nervosis lanceolatis* of 1758. William Fawcett and Alfred Rendle transferred it to *Eulophia* in 1910.

A large terrestrial herb 1–2 m tall. Perennating organ a fleshy elongate subterranean rhizome 1 cm diameter. Leaves 4, plicate-lanceolate, acuminate, up to 120 × 5–7.5 cm, long-petiolate. Inflorescence densely many-flowered; peduncle 8–10 mm diameter; rhachis

16–35 cm long. Bracts linear-aristate, 10–80 mm long. Flower not opening widely; sepals olive-green or green; lip and petals red-purple, rarely albino; pedicel and ovary 15–23 mm long. Dorsal sepal oblong-lanceolate, 18–20 × 4–5 mm; lateral sepals oblong or ligulate, 18–22 × 4–6.5 mm. Petals oblong-ligulate, 15–19 × 3–5 mm. Lip 3-lobed in apical half, saccate at base, 14–19 × 8–12 mm; side lobes rounded; midlobe semicircular, 5–6 × 8 mm, with an undulate margin; callus of 2 obliquely subquadrate flap-like ridges in middle of lip and with papillae on the central 5 veins of the midlobe.

E. alta

Habitat and distribution In papyrus and other swamps and in wet grassland between 1150 and 1300 m in Kefa. Also in Guinea, Sierra Leone, Liberia, Ivory Coast, Ghana, Nigeria, Cameroon, Gabon, Central African Republic, Sudan, DR Congo, Uganda, Angola, Zambia, Zimbabwe and tropical America.

Flowering period May.

Conservation status Rare in Ethiopia but otherwise widely distributed and often locally common.

Notes Readily recognised by its large stature, relatively small flowers with green sepals and petals and deep purple saccate lip.

Eulophia lips (flattened) and spurs

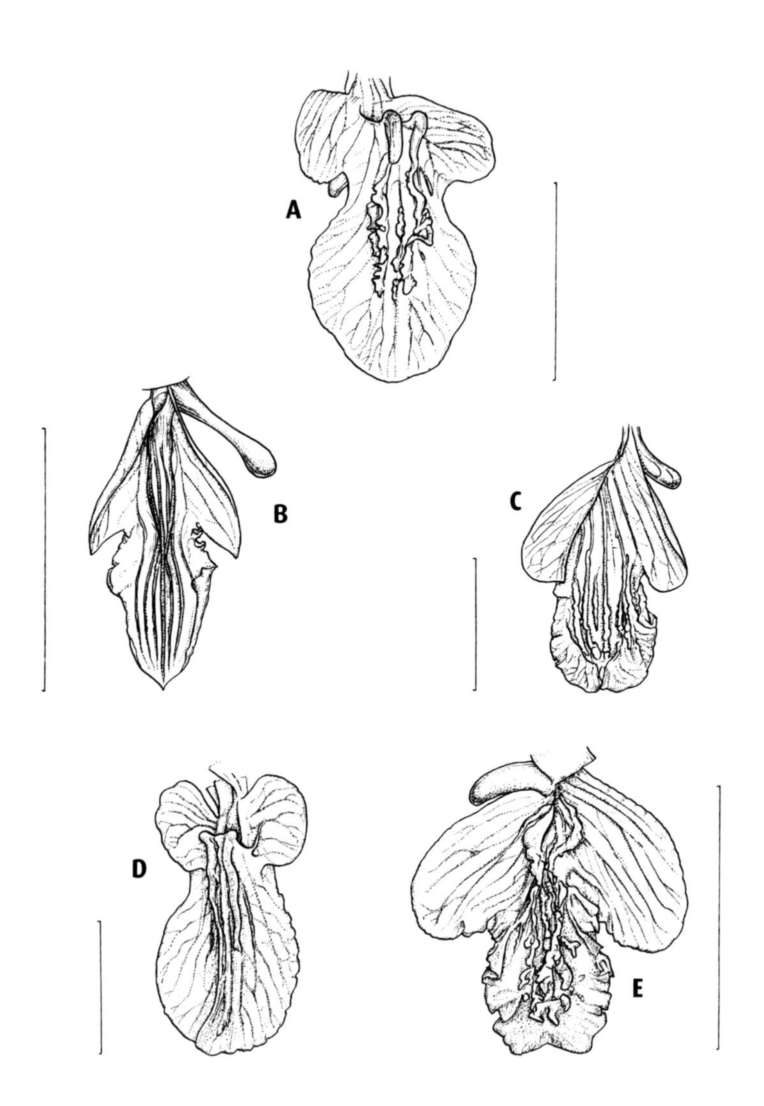

A *E. cristata;* **B** *E. euglossa;* **C** *E. flavopurpurea;* **D** *E. livingstoniana;* **E** *E. stachyodes.* Scale bar = 10 mm.

6. E. stachyodes *Rchb.f.*

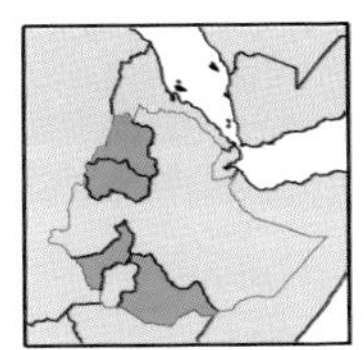
Eulophia stachyodes

The specific epithet '*stachyodes*' means like an ear of corn. H.G. Reichenbach described it in 1878 from a type specimen collected by Georg Schweinfurth at Niamniam, Sudan.

Plant 30–80 cm tall. Perennating organs underground, tuberous. Leaves 4–5, plicate, lanceolate or oblanceolate, 17–100 × 3–8.5 cm, long-petiolate. Inflorescence produced with the leaves, densely many-flowered; peduncle 3–5 mm diameter; rhachis 6–26 cm long. Bracts linear-lanceolate, 10–55 mm long. Flowers spreading; sepals dull green, bronze or greenish red; petals white, lip green with a purple callus; spur green flushed with purple; pedicel and ovary 15–16 mm long. Dorsal sepal narrowly oblong, 18–23 × 2.5–4.5 mm; lateral sepals similar. Petals elliptic-oblong, 16–18 × 6.5–8 mm. Lip 3-lobed, 14–15 × 9 mm; side lobes broadly oblong, rounded in front; midlobe oblong, 6 × 4.5 mm, undulate on margin; callus of 3 ridges with 3–5 rows of short papillae on midlobe; spur conical-cylindrical, 4 mm long.

Habitat and distribution In grassland, bushland and woodland on humid blackish soils between 950 and 2100 m in Gonder, Gojam, Sidamo and Kefa. Also in Uganda, Kenya, Tanzania, Nigeria to Sudan and south to Zimbabwe.

Flowering period June to August.

Conservation status Rare.

Notes Readily recognised by its spreading green and white flowers with violet markings in the lip.

E. stachyodes

7. E. adenoglossa (*Lindl.*) *Rchb.f.*

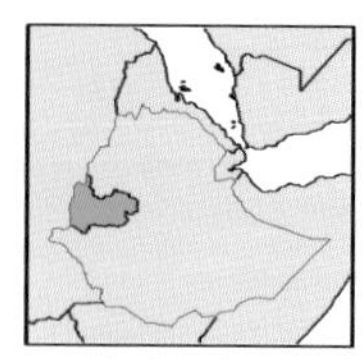

Eulophia adenoglossa

The specific epithet comes from the Greek *aden*, a gland, and *glossa*, a tongue, referring to the gland-bearing lip. John Lindley described it in 1878 as *Cyrtopera adenglossa* based upon a collection from Nigeria. H.G. Reichenbach transferred it to *Eulophia* in 1878.

Plant 45–95 cm tall. Tubers in chains, irregularly ovoid to ellipsoid, 3–5 × 2–4 cm. Leaves 2–3, erect, linear, acuminate, 18–40 × 0.5–2. Inflorescence erect, produced with the young leaves, laxly 7–17-flowered; peduncle bearing a few scattered sheaths; rhachis up to 25 cm long; bracts deflexed, aristate to linear-lanceolate, acuminate, 1–3.8 cm long. Flowers subnutant, waxy in texture, with green or yellow-green sepals and petals and a pale green to creamy green lip with purple-brown spots on the margin of the midlobe and yellow-green papillae; pedicel and ovary 1.2–1.6 cm long. Dorsal sepal lanceolate, acute, 15–26 × 4–6 mm. Lateral sepals similar but slightly longer and oblique at base. Petals narrowly oblong-lanceolate, acute, 15–21 × 4–6 mm. Lip 3-lobed in middle, 20–21 × 9–11 mm; side lobes erect, narrowly oblong, rounded in front; midlobe decurved, oblong-subquadrate, blunt or truncate; callus of two basal ridges to middle of lip, with several rows of papillae on midlobe; spur conical, 2–5 mm long. Column 8–9 mm long; foot short.

E. adenoglossa

Habitat and distribution In semi-open bamboo thicket at 1270 m in Wellega. Also in Ghana, Nigeria, Ethiopia, Kenya, Tanzania, Zambia, Zimbabwe, Malawi, Mozambique and South Africa.

Flowering period June and July.

Conservation status Very rare in Ethiopia but locally common elsewhere.

Notes Close to *E. stachyodes* but differs in its subnutant flowers with a pale green rather than white lip with a callus of 2 rather than 3 basal ridges.

8. E. horsfallii (*Bateman*) *Summerh.*

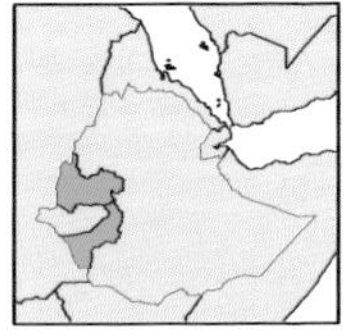

Eulophia horsfallii

The specific epithet '*horsfallii*' is in honour of J.B. Horsfall who first flowered the plant in England. James Bateman described it as *Lissochilus horsfallii* in 1865 from a plant collected by Cheetham by the Old Calabar River, Nigeria. Victor Sumerhayes transferred it to *Eulophia* in 1936.

A large terrestrial herb 1.2–3 m tall, growing from stout subterranean rhizomes, at least 4 cm long. Leaves 3–5, erect, lanceolate to oblanceolate, 30–140 × 1.6–15.5 cm, long-petiolate. Inflorescence erect, laxly 5–50-flowered; peduncle stout, up to 3.5 cm diameter at base; rhachis 16–37 cm long; bracts elliptic or obovate, 19–35 × 10–18 mm. Flowers large, fleshy; sepals dull purple or brown; petals rose-purple; lip side

E. horsfallii

lobes green striped with dull purple, midlobe purple; callus yellow; column and ovary purple; pedicel and ovary 30–40 mm long. Dorsal sepal oblanceolate, 17–26 × 7–8 mm; lateral sepals oblique at base but similar, 17–27 × 7–8 mm. Petals covering column, elliptic, 19–35 × 15–24 mm. Lip 3-lobed, 20–40 × 14–40 mm; side lobes erect, semicircular; midlobe elliptic-oblong, with undulate margins; callus in basal two-thirds of lip, 3-cristate, raised at base, undulate and lower in front; spur conical, 4–8 mm long, slightly retrorse.

Habitat and distribution In swampland at about 1530 m in Kefa and Wellega. Also throughout tropical Africa and south to South Africa.

Flowering period May and October.

Conservation status Rare and possibly endangered in Ethiopia but locally common elsewhere.

Notes Readily distinguished by its large flowers with deep maroon sepals, large, almost circular or elliptic purple petals and a purple lip with a short upturned spur at the base and a tricristate callus.

9. E. angolensis *(Rchb.f.) Summerh.*

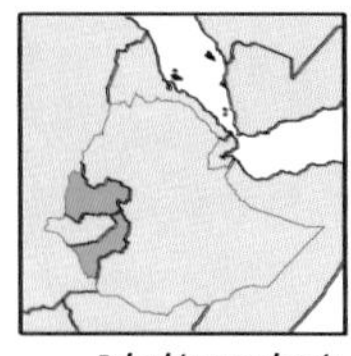

Eulophia angolensis

The specific epithet '*angolensis*' refers to Angola, where the type specimen was collected. H.G. Reichenbach described it as *Cymbidium angolense* from a plant collected by Friedrich Welwitsch in Huilla, Angola. Victor Summerhayes transferred it to *Eulophia* in 1958.

Plant 60–120 cm tall, growing from fleshy, subterranean many-noded rhizomes. Leaves 2–4, plicate, linear-lanceolate, 50–130 × 0.7–4.3 cm. Inflorescence laxly many-flowered, rarely 1-branched; peduncle stout, 1–2 cm diameter; rhachis 16–40 cm long. Bracts ovate, lanceolate or oblanceolate, 1–2.5 cm long. Flowers spreading or subnutant; sepals purple-brown or yellow; petals and lip yellow; pedicel and ovary 14–21 mm long. Dorsal sepal ligulate-spathulate, 16–27 × 5.3–7.5 mm; lateral sepals similar 16–25 × 5.3–7.5 mm. Petals parallel to and covering column, elliptic, rounded at apex, 16–25.5 × 9.5–13.8 mm. Lip 3-lobed, saccate at base, 16–25 × 10–16 mm; side lobes elliptic, rounded; midlobe oblong or obovate, undulate on margin; callus of 3 tall crenulate ridges, flexuous in front.

E. angolensis

Habitat and distribution In marshy meadows between 1310 and 2000 m in Kefa and Wellega. Also throughout tropical Africa.

Flowering period May to July.

Conservation status Vulnerable.

Notes *E. angolensis* differs from *E. horsfallii* by the petals and lip being yellow rather than purple.

10. E. caricifolia *(Rchb.f.) Summerh.*

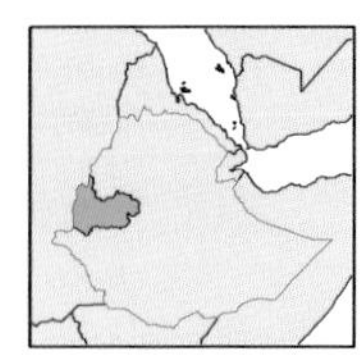
Eulophia caricifolia

The specific epithet 'caricifolia' refers to the sedge-like leaves. Described by H.G. Reichenbach as *Lissochilus caricifolius* in 1877 based on plants collected by Griffon de Bellay in Gabon. Victor Summerhayes transferred it to *Eulophia* in 1936.

Plant 35–80 cm tall. Rhizome elongate, cylindrical but swollen below stems, up to 1.2 cm in diameter. Stems short, leafy. Leaves 4–6, in a fan, suberect-arcuate, linear-lanceolate, acuminate, the largest 35–55 × 1.5–1.9 cm. Inflorescence erect, unbranched, laxly 7–15-flowered; bracts lanceolate, acuminate, 4–9 cm long. Flowers showy, sepals maroon-brown, petals pale pink, lip pink, callus on lip bright yellow; pedicel and ovary 1.6–2.7 cm long. Dorsal sepal erect, subspathulate, rounded and concave at apex, 1.5–1.7 × 3–4 mm. Lateral sepals suberect-reflexed, subspathulate, obtuse, 1.6–1.8 × 4–4.5 mm. Petals subporrect, elliptic, rounded at apex, 1.4–1.5 × 0.8–1 cm. Lip porrect, 3-lobed, angled to shortly conically spurred at base; side lobes elliptic, rounded in front; midlobe ovate-elliptic, obtuse, margin crispate-undulate; callus of 3 obscure fleshy crests which run from the base into several strongly papillate ridges extending almost to the apex of midlobe.

Habitat and distribution In swampy meadows by river at 1500 m in Wellega. Widespread in tropical Africa.

Flowering period May to August.

Notes Its closest ally is the widespread *E. angolensis*, which is readily distinguished by its stouter habit, broader leaves, yellow and maroon or pure yellow flowers and details of the callus of the lip.

E. caricifolia

11. E. flavopurpurea *(Rchb.f.) Rolfe*

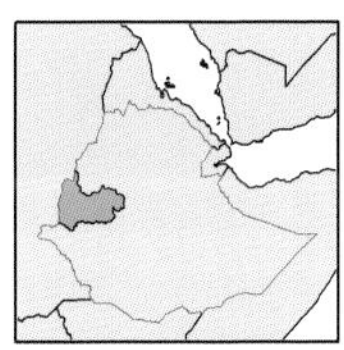

Eulophia flavopurpurea

The specific epithet '*flavopurpurea*' refers to the yellow and purple flowers. H.G. Reichenbach described it in 1878 as *Cyrtopera flavopurpurea*, based upon a collection made by Georg Schweinfurth in Niamniam, Sudan. Robert Rolfe transferred it to *Eulophia* in 1897.

Plant 60–95 cm tall, growing from underground irregularly shaped corms. Corms 3–4 × 1–2 cm, 2–3-noded, with scattered white roots. Leaves 3–4, erect, linear, 5–35 × 0.5–1.5 cm. Inflorescence produced before the leaves have developed, laxly few-flowered; peduncle up to 75 cm long, bearing 3 widely spaced sheaths; rhachis up to 20 cm long; bracts spreading, linear, acuminate, 1.5–3 cm long. Flowers pale greenish yellow, sometimes flushed purple on back of sepals, lip yellowish green or white, sometimes with purple venation on midlobe; pedicel and ovary 1.3–2.1 cm long. Dorsal sepal lanceolate, 2–3.4 × 0.4–0.7 cm. Lateral sepals obliquely lanceolate, 2.4–3.2 × 0.4–0.7 cm. Petals porrect, elliptic-ovate, 2–2.4 × 0.7–1.1 cm. Lip 3-lobed, 2.3–2.9 × 1–1.5 cm when flattened; side lobes narrowly oblong, rounded in front; midlobe deflexed, flabellate-obovate, retuse; callus of 5–7 low verrucose ridges along veins of midlobe; spur conical, 0.4–0.7 cm long.

E. flavopurpurea

Habitat and distribution In open meadows with scattered bushes and in open *Combretum* woodland between 1040 and 1530 m in Wellega. Widespread in West and East tropical Africa.

Flowering period May to July.

Notes Always flowers before the leaves develop.

12. E. cucullata *(Sw.) Steud.*

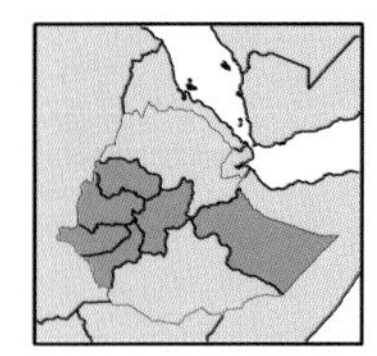

Eulophia cucullata

The specific epithet '*cucullata*' refers to the cucullate lip. Olof Swartz described it as *Limodorum cucullatum* in 1800, from a plant collected by Adam Afzelius in West Africa. Ernst Steudel transferred it to *Eulophia* in 1840.

Plant 40–130 cm tall, growing from irregularly fusiform-conical, potato-like tubers, 4–5.5 × 2–3 cm. Leaves 3–4, plicate, linear, 20–70 × 0.4–1.7 cm. Inflorescence produced before or with young leaves, laxly 3–8-flowered; peduncle stout, 3–12 mm diameter; rhachis 6–25 cm long. Bracts linear, 35–55 mm long. Flowers showy, variable in size; sepals maroon, brown or ochre; petals rose-purple; lip rose-purple with a white to cream throat speckled and streaked with purple, side lobes greenish striped with brown; pedicel and ovary 8–25 mm long. Dorsal sepal lanceolate, 14–24 × 3–8 mm; lateral sepals erect or reflexed, obliquely lanceolate, 14–25 × 3–10 mm. Petals parallel to column, elliptic or obovate, rounded, 15–24 × 10–21 mm. Lip 3-lobed, saccate, 18–34 × 20–40 mm wide; side lobes rounded; midlobe occasionally clawed, transversely oblong, emarginate, 11–12 × 16–22 mm; callus of 2 parallel quadrate ridges at base of midlobe

Habitat and distribution In *Combretum/Terminalia* savannah and grassland, in sandy soil between 550 and 2000 m in Gojam, Shewa, Harerge, Illubabor, Kefa and Wellega. Also throughout tropical and southern Africa and in Madagascar.

Flowering period May to August.

Conservation status Vulnerable Ethiopia but locally common elsewhere.

Notes Readily recognised by its large pink flowers with a saccate lip bearing two subquadrate yellow calli.

E. cucullata

13. E. cristata (*Sw.*) *Steud.*

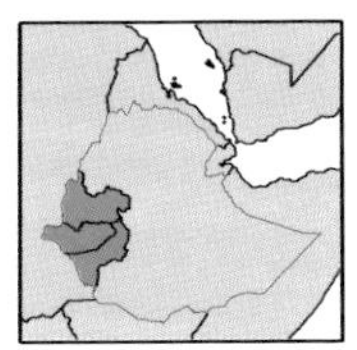

Eulophia cristata

The specific epithet '*cristata*' refers to crests of hairs on the lip. Olof Swartz described it as *Limodorum cristatum* in 1805 from a plant collected by Adam Afzelius in Sierra Leone. Ernst Steudel transferred it to *Eulophia* in 1840.

Plant 60–130 cm tall, growing from underground, potato-like, cylindric-ellipsoidal tubers, 3–9 × 1.3–4 cm, whitish, in chains. Leaves 4–6, erect, plicate, lanceolate, 40–70 × 0.7–3.5 cm, petiolate, basal 1–3 sheathing. Inflorescence appearing before leaves, laxly 10–30-flowered; peduncle 3–8 mm diameter; rhachis 13–44 cm long. Bracts aristate, 10–30 mm long. Flowers with lilac sepals and petals, a dark purple lip and greenish column; pedicel and ovary 15–27 mm long. Dorsal sepal oblanceolate to narrowly elliptic, 14–23 × 3–5.5 mm; lateral sepals similar but slightly larger and oblique at base. Petals elliptic, 12.5–19 × 6.5–9.5 mm. Lip 3-lobed, 12.5–22 × 10–13.5 mm; side lobes erect, semicircular, recurved; midlobe deflexed, somewhat convex, elliptic, crisped on margin; callus of 2 raised rounded ridges at base with 5–9 crenulate ridges in front on midlobe; spur slightly upcurved, conical, 2.5–5 mm long.

E. cristata

Habitat and distribution In seasonally burnt grassland, *Commiphora-Albizia* and *Entada-Acacia* bushland and woodland between 700 and 1850 m in Illubabor, Kefa and Wellega. Also in Senegal and Gambia across to Sudan and Uganda.

Flowering period February to June.

Conservation status Rare and vulnerable in Ethiopia but widespread and often common in West Africa.

Notes *E. cristata* differs from *E.livingstoniana* by the presence of cristate-papillose raised ridges on the midlobe of the lip.

14. E. livingstoniana *(Rchb.f.) Summerh.*

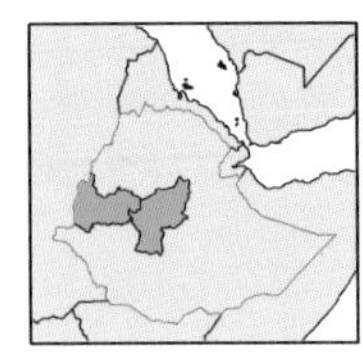

Eulophia livingstoniana

The specific epithet '*livingstoniana*' commemorates David Livingstone, who led the Zambesi expedition on which it was discovered. H.G. Reichenbach described it as *Lissochilus livingstonianus* in 1881 based on collections by Charles Meller and Horace Waller in Manganja Hills, Malawi. Victor Summerhayes transferred it to *Eulophia* in 1948.

Plant 50–100 cm tall, growing from irregularly cylindrical underground tubers. Leaves 3–6, linear, 10–40 × 5–17 mm. Inflorescence produced before leaves develop, laxly few–many-flowered; peduncle 37–75 cm long; rhachis 10–26 cm long. Bracts linear-lanceolate, lowest 17–19 mm long. Flowers with pale to darker lilac sepals and petals; lip purple with darker callus ridges and green or yellowish side lobes flushed and edged with pink; pedicel and ovary 22–30 mm long. Dorsal sepal lanceolate or oblanceolate, 15–20 × 4–5 mm; lateral sepals similar but slightly wider. Petals erect, elliptic-obovate, 14–20 × 8–9 mm. Lip 3-lobed, 12–17 mm long; side lobes erect, rounded; midlobe reduplicate, elliptic, 8–12 mm long; callus of 5 low ridges to apex of midlobe; spur narrowly conical, 3–7 mm long, slightly upcurved at apex.

E. livingstoniana

Habitat and distribution In alluvial soil by rivers and in swampy meadows between 1200 and 1430 m in Shewa and Wellega. Also widespread in tropical East and South-central Africa from Sudan south to Botswana and in Madagascar.

Flowering period January to May.

Conservation status Rare in Ethiopia but locally frequent elsewhere.

Notes *E.livingstoniana* differs from *E. cristata* by the midlobe of the lip being smooth rather than ridged.

Eulophia lips (flattened) and spurs

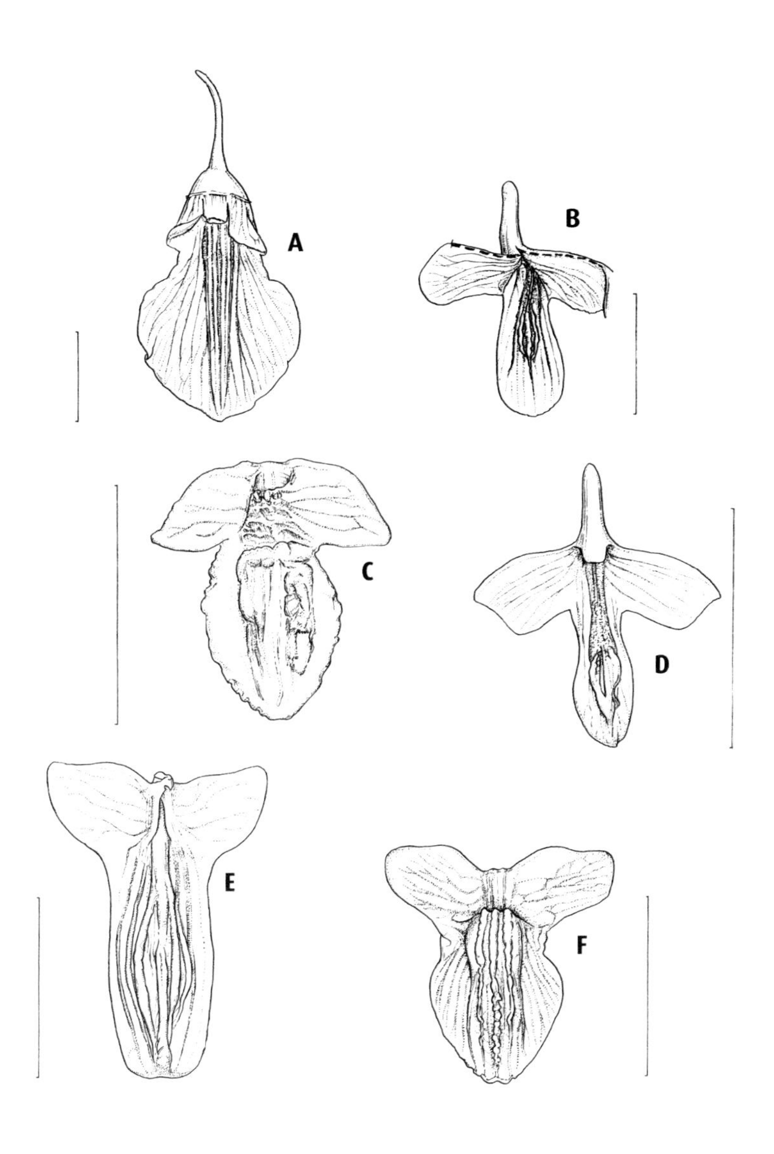

A *E. guineensis;* **B** *E. orthoplectra;* **C** *E. pyrophila;* **D** *E. schweinfurthii;* **E** *E. speciosa;* **F** *E. streptopetala.*
A, B, D–F Scale bar = 10 mm; C Scale bar = 5 mm.

15. E. speciosa *(Lindl.) Bolus*

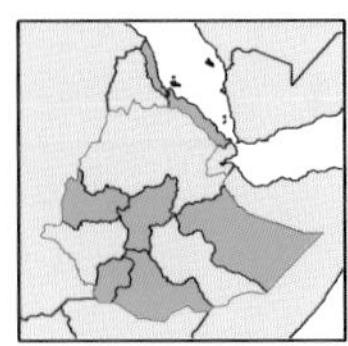

Eulophia speciosa

The specific epithet '*speciosa*' refers to the beautiful flowers. John Lindley described it as *Lissochilus speciosus* in 1821 from a plant collected in South Africa and cultivated by Griffin. Harry Bolus transferred it to *Eulophia* in 1889.

Plant 60–120 cm tall, growing from irregularly conical underground tubers, 4–6 × 2.5–4 cm. Leaves 3–6, in a fan, 30–65 × 0.6–2.2 cm. Inflorescence produced before the leaves, laxly many-flowered; peduncle stout, purplish; rhachis 11–29 cm long. Bracts lanceolate, 4–14 mm long. Flowers variable in size; sepals greenish; petals yellow; lip yellow with red veins on side lobes; column pale greenish; pedicel and ovary 12–19 mm long. Dorsal sepal reflexed, elliptic, 5–11 × 2–4 mm; lateral sepals similar. Petals spreading, elliptic-ovate, 11–18 × 8–15 mm. Lip 3-lobed, 12–16 × 10–14 mm; side lobes obscurely rounded; midlobe convex, elliptic, 10–13 × 8–12 mm; callus of 3–5 low fleshy rugulose ridges on midlobe; spur shortly conical, 1–3 mm long.

Habitat and distribution In grassland, deciduous bushland and woodland, in shade of *Euphorbia* species between 1200 and 2000 m in Shewa, Harerge, Sidamo and Gamo Gofa. Also in SW Arabia, Uganda, Kenya, Tanzania, Zanzibar, Sudan south to Zimbabwe and South Africa.

Flowering period March to June.

Conservation status Vulnerable but locally common in Ethiopia.

Notes *E. speciosa* differs from *E.orthoplectra* and *E. schweinfurthii* in having pure yellow flowers slightly marked with red streaks on the lip and by its shorter spur.

E. speciosa

16. E. orthoplectra *(Rchb.f.) Summerh.*

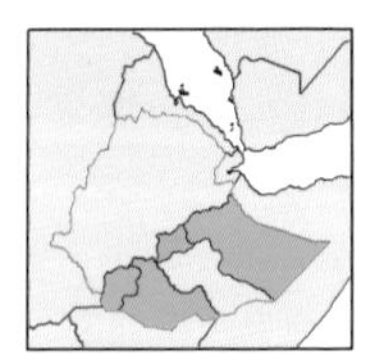

Eulophia orthoplectra

The specific epithet '*orthoplectra*' refers to the straight spur. H.G. Reichenbach described it as *Lissochilus orthoplectrus* in 1878 from a plant collected by Georg Schweinfurth at Niamniam, Sudan. Victor Summerhayes transferred it to *Eulophia* in 1939.

Plant 60–100 cm tall, growing from irregularly ovoid-fusiform underground tubers, 3.5–6 × 2–5 cm. Leaves 2–4, fleshy, linear, 18–50 × 0.7–2 cm. Inflorescence laxly 6–20-flowered, produced before leaves; peduncle with 4–6 sheaths; rhachis 9–20 cm long. Bracts linear-lanceolate, 8–12 mm long. Flowers with red-brown sepals; petals yellow, red-brown within and red-veined; lip yellow with red-veined side lobes and a brown spur; pedicel and ovary 10–13 mm long. Dorsal sepal oblanceolate, 8.5–12.5 × 3.8–6.3 mm; lateral sepals similar. Petals elliptic-subcircular or elliptic-ovate, 12–17.5 × 13–19 mm. Lip 3-lobed, 11–15 mm long; side lobes rounded; midlobe convex, elliptic-obovate; callus of 5 fleshy rugulose ridges; spur conical-cylindrical with a broad mouth, 13–16 mm long.

Habitat and distribution In grassland, wooded grassland, swamp and *Brachystegia* woodland between 1100 and 2100 m in Harerge, Sidamo and Gamo Gofa. Also in Nigeria and Cameroon to Sudan, DR Congo, Rwanda, Uganda, Kenya, Tanzania, Malawi, Zambia, Mozambique and Zimbabwe.

Flowering period May to August.

Conservation status Rare and vulnerable.

Notes *E. orthoplectra* differs from *E. schweinfurthii* by the spur being 13–16 mm long rather than 8–9 mm long.

E. orthoplectra

17. E. schweinfurthii *Kraenzl.*

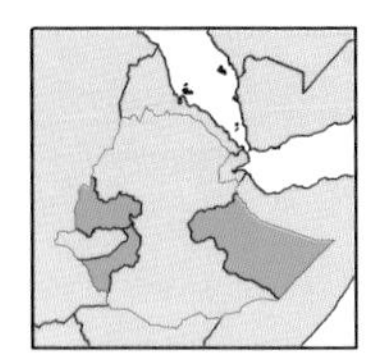
Eulophia schweinfurthii

The specific epithet '*schweinfurthii*' was given in honour of the German collector Georg Schweinfurth who collected the type in Bongoland, Sudan. Fritz Kraenzlin described it in 1893.

Plant 45–110 cm tall, growing from irregularly cylindrical underground tubers, in chains, 3.5–5 × 1.3–2.5 cm. Leaves 4–7, linear, 21–38 × 2.5–10 mm. Inflorescences 1–2, produced before the leaves; peduncle up to 7 mm diameter; rhachis 7–23 cm long. Bracts linear-lanceolate, 4–15 mm long. Flowers fleshy; sepals purplish; petals yellow with red veins; lip yellow with pale purplish side lobes and a purplish margin to the midlobe; pedicel and ovary 10–20 mm long. Dorsal sepal oblong, 9–10.6 × 3.5–6 mm; lateral sepals similar. Petals broadly ovate or subcircular, 11.5–16.5 × 9.5–17 mm. Lip 3-lobed, 14–16 × 19–21 mm; side lobes rounded; midlobe convex, elliptic, 7–12 mm long; callus of 3–5 verrucose ridges to apex on midlobe; spur upcurved, conical, 8–9 mm long.

Habitat and distribution In bushland with *Acacia* and *Clerodendrum*, wooded grassland with *Terminalia* and *Grewia,* stony ground, limestone or sandstone between 1200 and 1800 m in Harerge, Kefa and Wellega. Also in Uganda, Kenya, Tanzania, Sudan, DR Congo, Angola, Zambia, Malawi, Mozambique, Zimbabwe and Botswana.

Flowering period March and April; September to December.

Conservation status Rare.

Notes *E. schweinfurthii* differs from *E.orthoplectra* by the spurs being 8–9 mm rather than 13–16 mm long.

18. E. pyrophila *(Rchb.f.) Summerh.*

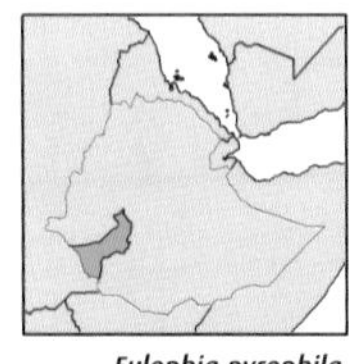
Eulophia pyrophila

The specific epithet '*pyrophila*', meaning fire-lover, refers to the fact that this species often flowers after vegetation has been burnt. H.G. Reichenbach described it as *Lissochilus pyrophilus* in 1878, from a plant collected by Georg Schweinfurth near Kuraggera, Sudan. Victor Summerhayes transferred it to *Eulophia* in 1948.

Plant 13–45 cm tall, growing from irregularly conical underground tubers, 1–3 × 1–2.5 cm. Leaves 4–5, linear-lanceolate, 3–14 × 2–2.5 mm. Inflorescence produced before the leaves, laxly many-flowered; rhachis 2–3 mm diameter. Bracts linear, 3–9 mm long. Flowers chocolate or dull brown with a cream or yellow lip striped brown and with a yellow callus; pedicel and ovary 9–20 mm long, plum-coloured. Dorsal sepal oblong-elliptic, 5–6.7 × 2.5–3.6 mm; lateral sepals similar. Petals elliptic or oblong-elliptic, 5.8–8 × 4.3–5.7 mm. Lip 3-lobed, 5.7–7.5 × 5–6.8 mm; side lobes triangular, united for half their length to the column; midlobe convex, elliptic, margin undulate; callus of 5–9 fleshy rugulose ridges on basal half of midlobe; spur conical, 1.5–2.5 mm long.

Habitat and distribution In short, often burnt, grassland and amongst rocks at about 1600 m in Kefa. Also in Uganda, Kenya, Tanzania, Ivory Coast and south to Zimbabwe.

Flowering period January.

Conservation status Rare.

Notes *E. pyrophila* differs from other species in the genus with subcircular yellow petals by having smaller flowers with 5–7 mm long sepals and petals.

E. pyrophila

19. E. abyssinica *Rchb.f.*

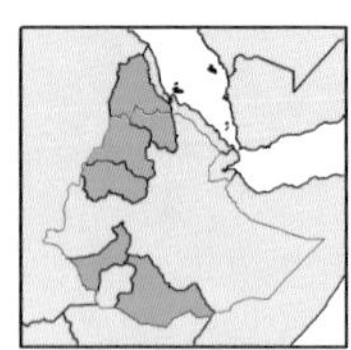

Eulophia abyssinica

The specific epithet '*abyssinica*' refers to Abyssinia, the old name for Ethiopia. H.G. Reichenbach described it in 1850 from a plant collected by Georg Wilhelm Schimper near Adde Schum Eschet in Tigray.

Plant growing from irregularly shaped underground tubers. Leaves plicate, lanceolate, up to 40 × 5.5 cm. Inflorescence up to 75 cm tall, densely many-flowered; peduncle almost covered by paper-like tubular sheaths. Bracts lanceolate or linear, 1–3 cm long. Flowers spreading, not opening widely, mustard yellow, sometimes with an orange mark on lip; pedicel and ovary 10–14 mm long. Dorsal sepal lanceolate, 20–25 × 6–8 mm. Lateral sepals slightly obliquely lanceolate, 21–30 × 6–8 mm. Petals lanceolate, 15–25 × 7–8 mm. Lip 3-lobed in middle, 14.5–18 × 9–10 mm; side lobes rounded in front; midlobe broadly ovate to subcircular, half length of lip, glabrous; callus of 2 low papillose ridges in basal half of lip; spur conical, 3–4 mm long.

Habitat and distribution In moorland, montane pasture and on shrubby hillsides between 2250 and 2600 m in Tigray, Gonder, Gojam, Sidamo and Kefa and in Eritrea. Unknown elsewhere.

Flowering period May to July.

Conservation status Vulnerable.

Notes *E. abyssinica* differs from *E. zeyheri* by the midlobe of the lip lacking papillae, hairs, and maroon markings.

20. E. albobrunnea *Kraenzl.*

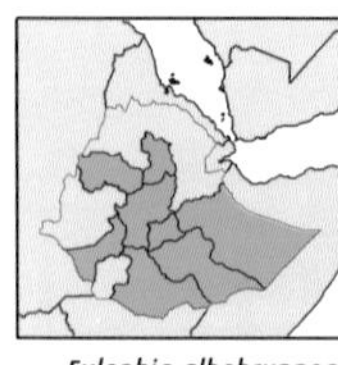

Eulophia albobrunnea

The specific epithet '*albobrunnea*' refers to the flower colour which is white to pale pink or lilac with a dark purple midlobe. Fritz Kraenzlin described it in 1902 from a plant collected by Hans Ellenbeck in Diddah, Bale/Harerge.

Terrestrial herb with underground subglobose to cubical, rhizome-like, white tubers, 2–2.5 cm long. Leaves 2, erect, plicate, linear-lanceolate, 9–70 × 1.5–2.6 cm, glaucous or pale green with yellow-green venation. Inflorescence 20–70 cm tall; peduncle covered by whitish paper-like sheaths, turning brown with age. Bracts lanceolate, up to 25 × 5 mm. Flowers waxy, somewhat fragrant, white to pale pink or lilac with a dark purple midlobe and papillae on the lip; pedicel and

Eulophia lips (flattened) and spurs

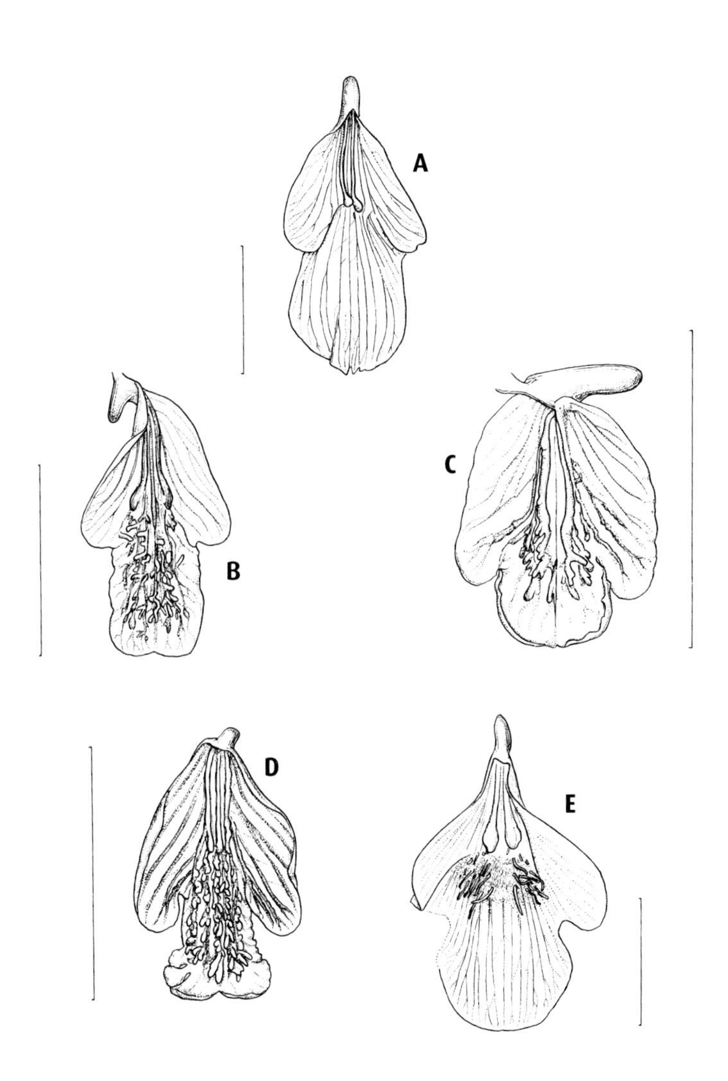

A *E. abyssinica;* **B** *E. adenoglossa;* **C** *E. clavicornis* var. ***nutans;***
D *E. kyimbilae;* **E** *E. zeyheri.* Scale bar = 10 mm.

ovary 8–9 mm long, glabrous. Dorsal sepal oblong-ovate, 11–12 × 4–5 mm. Lateral sepals oblong or oblong-elliptic, 10–12.5 × 4.5–5 mm. Petals oblong, 10.5–12.5 × 4.5–5 mm, margins somewhat undulate. Lip 3-lobed in apical half, 7–9 × 7–7.5 mm; side lobes semielliptic; midlobe subcircular, 3 × 2.5–3 mm, densely longly papillate; spur very obscure, less than 1 mm long.

Habitat and distribution In montane grassland with scattered trees, wet meadows and rough montane grassland between 1650 and 2500 m in Welo, Gojam, Shewa, Arsi, Harerge, Bale, Sidamo and Kefa. Unknown elsewhere.

Flowering period June to October.

Conservation status Vulnerable.

Notes *E. albobrunnea* differs from the other species in the genus by the lip lacking a column-foot and by the absence of a distinct spur or saccate base to the lip.

21. E. milnei *Rchb.f.*

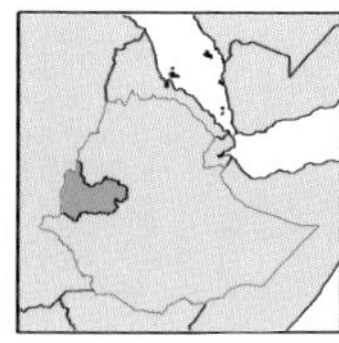

Eulophia milnei

The specific epithet '*milnei*' commemorates William Milne who discovered the plant. H.G. Reichenbach described it in 1881 based on a collection from Fernando Po.

Plant with underground tuberous ovoid perennating organs, 1–1.3 × 1 cm, borne in strings. Leaves 2–6, erect, linear, grass-like, 10–30 × 0.1–0.2 cm. Inflorescence produced before the leaves develop, densely many-flowered, peduncle almost covered by 5 papery sheaths; rhachis 0.5–1 cm long; bracts linear, acuminate, 1 cm long. Flowers very small, yellow, sometimes with reddish tinged side lobes to the lip; pedicel and ovary 0.5–0.8 cm long. Dorsal sepal oblong-lanceolate, acute to obtuse, 5–8.5 × 2.3–3 mm. Lateral sepals similar but slightly longer. Petals oblanceolate, subacute or acute, 5–8 × 2.5–3 mm. Lip 3-lobed, 5–7 × 3.5–4.5 mm; side lobes narrowly oblong, obtuse or rounded in front; midlobe oblong-obovate, emarginate or obtuse; callus of 3–5 lines of papillae on veins in apical two-thirds of lip; spur very short, cylindrical, 2.5–3.5 mm long.

Habitat and distribution In swampy meadows and seasonally wet grassland between 1280 and 1400m in Wellega. Also widespread in tropical Africa.

Flowering period May and June.

E. milnei

Notes Readily identified, having the smallest flowers (rarely more than 8 mm long including the spur) of any Ethiopian grassland *Eulophia* species.

22. E. odontoglossa *Rchb.f.*

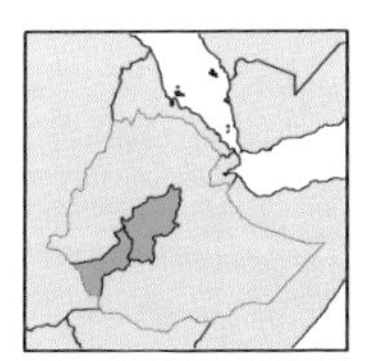

Eulophia odontoglossa

The specific epithet '*odontoglossa*' refers to the papillate, tongue-shaped lip. H.G. Reichenbach described it in 1847 from a plant collected in Natal, South Africa by W. Gueinzius.

Plant 60–100 cm tall. Perennating organs underground, tuberous, irregularly fusiform-conical or subglobose, 2–4 × 1.8–2.5 cm. Leaves 5–6, erect, plicate, oblanceolate, 40–70 × 1–2.1 cm. Inflorescence densely many-flowered; peduncle covered by 7 papery pale brown sheaths; rhachis 4–13 cm long. Bracts linear-aristate, 8–22 mm long. Flowers yellow with yellow, orange or red papillae on lip, less commonly brown or crimson; pedicel and ovary 12–21 mm long. Dorsal sepal ovate-elliptic, 9.3–12 × 4–5.8 mm; lateral sepals obliquely ovate, 9–14 × 3.8–5.2 mm. Petals obliquely elliptic, 8–12 mm × 3.3–5 mm. Lip 3-lobed, 8–11.5 × 4–8 mm; side lobes porrect; midlobe oblong, subquadrate or obovate; callus of 2 basal ridges with long papillae over basal three-quarters of midlobe; spur shortly conical-cylindric, 1–3 mm long.

Eulophia lips (flattened) and spurs

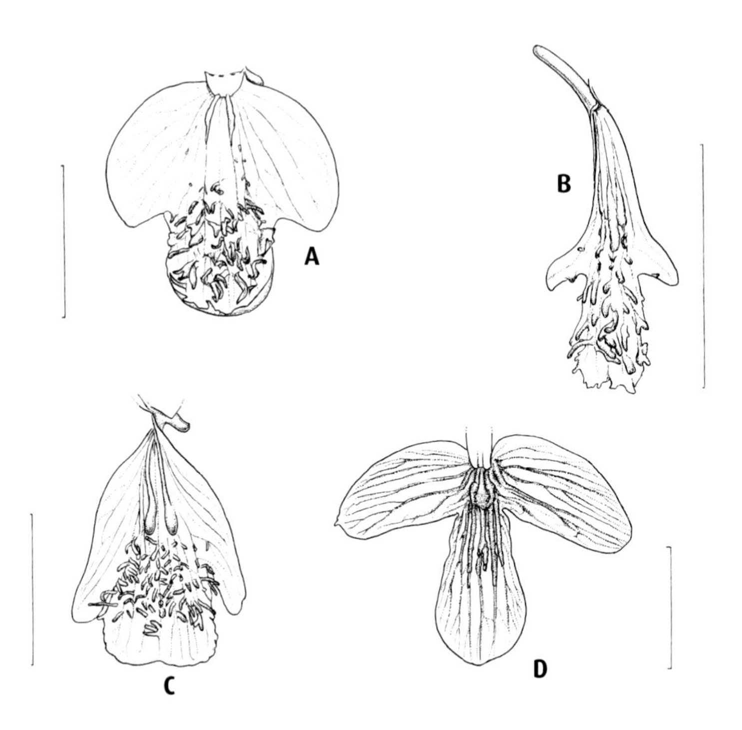

A *E. albobrunnea;* **B** *E. milnei;* **C** *E. odontoglossa;*
D *Pteroglossaspis eustachya.* Scale bar = 5 mm.

E. odontoglossa

Habitat and distribution In grassland and bushland, and in rocky areas between 1800 and 2350 m in Shewa and Kefa. Also in Uganda, Kenya, Tanzania, throughout tropical Africa from Guinea and Sierra Leone south to South Africa (Transvaal and Natal).

Flowering period June to August.

Conservation status Vulnerable in Ethiopia but locally common elsewhere in its range.

Notes *E. odontoglossa* differs from *E. milnei* in having larger flowers with red papillae on the lip.

23. E. zeyheri *Hook.f.*

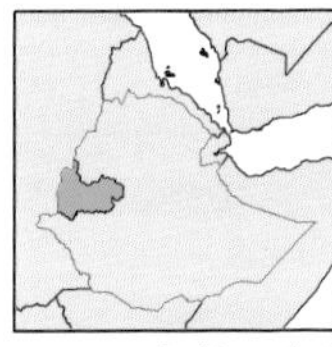

Eulophia zeyheri

Named for Carl Zeyher who collected one of the types in South Africa. Sir Joseph Hooker described it in 1893.

Plant up to 70 cm tall, growing from subterranean irregular tubers, 5–10 × 2–2.5 cm, borne in a string. Leaves erect, 2–3, lanceolate, acuminate, 30–50 × 1.3–4 cm. Inflorescence produced before or with the young leaves, subcapitate, densely many-flowered; peduncle covered by 3–4 slightly inflated sheaths; rhachis short; bracts subulate, 2.5–4 cm long. Flowers yellow, marked with maroon on the lip base and side lobes; pedicel and ovary 1.4–1.9 cm long. Dorsal sepal elliptic-lanceolate, acute, 28–36 × 12–16 mm. Lateral sepals similar but oblique at base. Petals elliptic-lanceolate, acute, 28–32 × 12–14 mm. Lip 3-lobed, 23–26 × 12–20 mm; side lobes erect, rounded in front; midlobe almost circular, obtuse to emarginate; callus of 2 ridges in basal half of the lip, with long papillose hairs at the apex.

Habitat and distribution In *Combretum-Terminalia* woodland and bushland between 1370 and 560 m in Wellega. Widespread in tropical and South Africa.

Flowering period May.

Notes Often confused with *E. abyssinica*. However, the lip of *E. abyssinica* is glabrous and the flowers are pale yellow, lacking maroon markings on the lip. *E. zeyheri* has a lip with long papillae or hairs at the apex of the column and base of the midlobe and is usually heavily marked with maroon on the base of the lip.

E. zeyheri

24. E. clavicornis *Lindl.*

The specific epithet '*clavicornis*' refers to the club-shaped spur. John Lindley described it in 1837 from a plant collected by Johann Drège in Katberg, South Africa.

E. clavicornis var. ***nutans***

var. **nutans** (*Sond.*) *A.V.Hall*

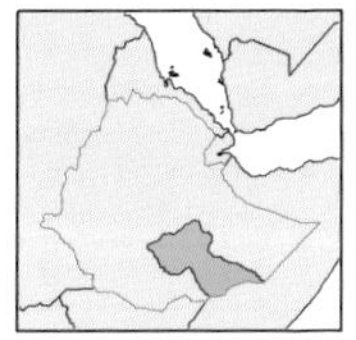

Eulophia clavicornis var. ***nutans***

The subspecific epithet '*nutans*' refers to the nodding flowers. Otto Sonder described it as *E. nutans* in 1846 from plants collected by Christian Ecklon and Carl Zeyher in Uitenhage and Katriviersberg, South Africa. Anthony Hall reduced it to varietal status in *E. clavicornis* in 1965.

Plant 26–45 cm tall, growing from underground tubers, 1.2–1.5 × 2 cm. Leaves 5–6, in a fan, plicate, linear, 17–40 × 3–8 mm. Inflorescence produced with leaves, laxly few–many-flowered; peduncle 2–3 mm diameter; rhachis 6–16 cm long. Bracts spreading, lanceolate, 9–18 mm long. Flowers rather small, often self-pollinating; sepals brown or green flushed with brown; petals white flushed with pink; lip white or yellowish flushed dull red-purple on edges; callus yellowish or greenish; pedicel and ovary 10–14 mm long. Dorsal sepal oblong-elliptic, 10–11 × 3–4 mm, keeled on outer side; lateral sepals similar but oblique, 11–12 × 4–4.5 mm. Petals elliptic, 8.5–10 × 3.5–5 mm. Lip 3-lobed, 7.5–10 × 5–9 mm; side lobes rounded in front; midlobe subcircular-oblong, 3 × 4.5 mm; callus of 2 ridges below and 4 lines of papillae on midlobe, all papillate; spur cylindrical-clavate, 2–3 mm long.

Habitat and distribution In grassland and bushland between 1650 and 2450 m in Bale. Also in Yemen, Kenya, Tanzania, Madagascar, Malawi, Zambia, Zimbabwe and South Africa.

Flowering period October.

Conservation status Rare and possibly threatened in Ethiopia but widespread and locally common throughout its range

Notes *E. clavicornis* differs from *E. stachyodes* in having smaller flowers.

25. E. kyimbilae *Schltr.*

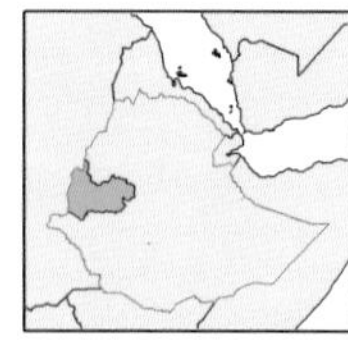

Eulophia kyimbilae

The specific epithet '*kymbilae*' refers to Kyimbila, on the northern tip of Lake Nyasa in Tanzania, where it was discovered by Adolf Stolz. Rudolf Schlechter described it in 1915.

Plant 50–80 cm tall. Pseudobulbs underground, tuber-like, irregularly cylindric-ellipsoidal, rooting on surface from nodes. Stems 2–3-leaved. Leaves erect, grass-like, linear-lanceolate, acuminate, 15–35 × 0.4–0.6 cm. Inflorescence erect, distantly laxly 7–9-flowered; bracts scarious, lanceolate, acuminate, 6–14 mm long. Flowers spreading to suberect, sepals and petals green with brownish purple venation, lip with white callus ridges and papillae on midlobe; pedicel and ovary 12–20 mm long. Dorsal sepal lanceolate, acute, 10–11 × 1.5 mm. Lateral sepals obliquely lanceolate, acute, 11–12 × 2 mm. Petals lanceolate, acute, 9–10 × 2.5 mm. Lip slightly curved in side view, more or less porrect, 3-lobed, 10–11 × 7 mm; side lobes narrowly oblong, rounded in front; midlobe obovate, slightly retuse, with strongly undulate erose margins, papillate; callus of 3 fleshy low ridges in basal half of lip terminating in several rows of short fleshy papillae; spur incurved-pendent, short, cylindrical, 1.5 mm long.

Habitat and distribution In swampy meadow by river between 1450 and 1500 m in Wellega. Also in Uganda, Kenya, Tanzania, Malawi and Zambia.

Flowering period July and August.

Notes Closely allied to the widespread *E. stachyodes*, also in Ethiopia, but which has much larger flowers with white petals and much broader leaves.

E. kyimbilae

26. GRAPHORKIS *Thouars.*

Epiphytic, sympodial herbs with clustered, cylindrical-fusiform, conical-ovoid or ovoid, several-noded pseudobulbs, partly covered by fibrous persistent leaf bases, leafy towards apex. Roots of two types: long, spreading ones attached to the substrate and erect tapering acuminate ones clustered around the base of the pseudobulb. Leaves plicate, suberect to erect, narrowly elliptic, acute or acuminate, shortly petiolate. Inflorescence proteranthous, erect, paniculate, many-flowered. Flowers small to medium, resupinate, yellow marked with brown or purple on the sepals and petals. Sepals and petals free, spreading. Sepals spathulate to oblong-elliptic, subacute. Petals narrowly elliptic. Lip 3-lobed, spurred at base; disc with keels; midlobe crenulate to bifid; spur often bent forward. Column with hirsute basal auricles; rostellum elongate; pollinia 2, waxy, sessile, attached to a solitary viscidium.

A genus of five species in Madagascar and the Mascarene Islands. A single species is found in tropical Africa and also occurs in Ethiopia.

G. lurida (*Sw.*) *Kuntze*

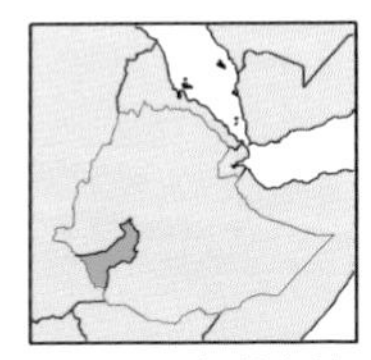

Graphorkis lurida

The specific epithet '*lurida*' refers to the purple or brown colour of the flowers. Olof Swartz described it as *Limodorum luridum* in 1805 from a plant collected by Adam Afzelius in Sierra Leone. Otto Kuntze transferred it to *Graphorkis* in 1891.

Pseudobulbs clustered, cylindrical-fusiform or conical-ovoid, 3–5-noded, 3–9 × 1–3 cm, 4–6-leaved. Leaves narrowly elliptic, 20–40 × 1.5–4.5 cm; petiole 2–4 cm long. Inflorescence paniculate, 15–50 cm long; branches 3–22 cm long. Bracts ovate-elliptic or narrowly elliptic, 8–25 mm long. Flowers with purple or brown sepals with pale green inner surfaces, pale green or cream petals flushed with brown or purple, and a yellow lip with greenish side lobes striped with brown; pedicel and ovary slender, 10–15 mm long. Sepals oblong-spathulate, 5–6.5 × 1.5–2 mm; petals elliptic, 4–6 × 2.5–3 mm. Lip 3-lobed, 5–6 × 3 mm; side lobes erect, oblong; midlobe obovate; callus of 2 fleshy keels on a disc; spur cylindrical, 3–4 mm long.

G. lurida

Habitat and distribution Found as an epiphyte on *Ficus lutea* and other canopy trees in forests of *Aningeria altissima*, *Morus mesozygia* and *Antiaris toxicaria* between 1100 and 1200 m in Kefa. Also in Senegal, Sierra Leone and Guinea across to Uganda and Tanzania and south to DR Congo and Burundi.

Flowering period December and January.

Conservation status Rare and vulnerable.

27. CALYPTROCHILUM *Kraenzl.*

Epiphytic, entirely glabrous monopodial herbs. Stems elongated, pendulous. Leaves fleshy, unequally bilobed at apex, alternate, twisted at base to lie in a parallel plane. Inflorescence abbreviated, lax or dense axillary racemes hidden beneath the leaves; bracts conspicuous or inconspicuous. Flowers resupinate, medium-sized. Tepals free, similar, apiculate, spreading, the petals slightly shorter. Lip 3-lobed, with a curved or geniculate spur which is inflated at the apex. Column short, with a small foot; rostellum prominent; anther apiculate; pollinia 2, globose, united to a long linear caudicle; viscidium solitary, large, triangular, grooved at the base to clasp rostellum.

A genus of two species in tropical Africa, extending as far south-east as Zimbabwe. Represented by a single species in Ethiopia.

C. christyanum *(Rchb.f.) Summerh.*

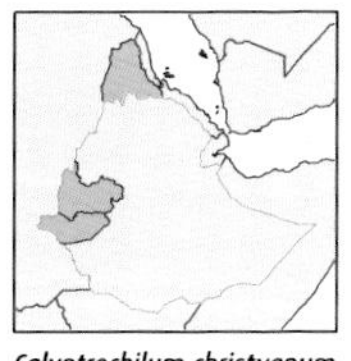
Calyptrochilum christyanum

The specific epithet '*christyanum*' was given in honour of Thomas Christy who collected the type specimen. H.G. Reichenbach described it as *Angraecum christyanum* in 1880 from a plant collected in West Africa. Victor Summerhayes transferred it to the present genus in 1936.

Stem simple, woody, rooting at nodes, leafy, 14–50 cm long. Leaves ligulate, unequally bilobed, born in one plane in 2 alternating rows, 4–13 × 0.8–2.5 cm. Inflorescences much shorter than leaves, 3–12-flowered; peduncle and rhachis 1.5–4 cm long. Bracts ovate, 2–4 mm long. Flowers white or greenish white, fading to apricot; base of lip yellow or green; spur yellow at base or entirely green; pedicel and ovary

C. christyanum

0.8–1 cm long. Dorsal sepal ovate to ovate-elliptic, 5–10 × 2.5–4 mm; laterals similar but slightly oblique. Petals oblong-elliptic, 5–8 × 2–3 mm. Lip distinctly 3-lobed, 7–12 × 7–10 mm; side lobes short, rounded, 2.5–5 × 3–5 mm; midlobe oblong or more or less rectangular, 5–7 mm long and broad; spur conical, broad below mouth, constricted in middle, inflated distally, 9–11 mm long.

Habitat and distribution In riverine forest and wooded grassland between 650 and 1400 m in Illubabor and Wellega. Also in Eritrea, Gambia, Ghana, Guinea, Guinea Bissau, Ivory Coast, Liberia, Mali, Nigeria, Sierra Leone, Cameroon, Central African Republic, DR Congo, Sudan, Kenya, Tanzania, Uganda, Angola, Malawi, Mozambique, Zambia and Zimbabwe.

Flowering period April and May.

Conservation status Rare in Ethiopia but widespread and locally common elsewhere.

C. christyanum

28. ANGRAECUM *Bory*

Dwarf to large epiphytic, lithophytic or rarely terrestrial herbs. Stems short to elongate, unbranched or branching, erect to pendent, covered by leaf-bases. Leaves thin-textured, fleshy or coriaceous, flattened or rarely iridiform, unequally bilobed at the apex, often twisted at the base and articulated to persistent leaf-bases. Inflorescences axillary, 1–many-flowered, racemose or rarely paniculate. Flowers fleshy, small to large, often stellate; ovary often twisted through 180 degrees, sessile or pedicellate. Sepals and petals free, subsimilar. Lip concave, entire to 3-lobed, enveloping the column at the base, spurred. Column fleshy, very short, lacking a foot; clinandrium rather shallow, cleft in front with a short tooth-like rostellum in the sinus; pollinia 2, globose, sulcate, attached to a common viscidium or each attached to separate viscidia.

A genus of about 200 species from tropical Africa, Madagascar and the adjacent islands. Three species are found in Ethiopia.

Key

1 Plants with elongate stems up to 60 cm long; leaves well spaced along stem 11–21.5 × 1.5–3 cm; flowers large; lip funnel-shaped, 6.5–8.5 cm long with a long slender apicule; spur geniculate, 16–21 cm long — **1. A. infundibulare**

\- Plants dwarf, stems less than 3 cm long; leaves clustered on stem; flowers small; lip less than 2 mm long; spur less than 2 mm long — 2

2 Leaves equitant (bilaterally flattened), 6–7 mm long, arranged along a short stem; spur clavate 1.4–1.8 mm long — **3. A. humile**

\- Leaves not equitant, 1.4–5.2 mm long, arranged in a fan; spur conical, 0.6 mm long — **2. A. minus**

1. A. infundibulare *Lindl.*

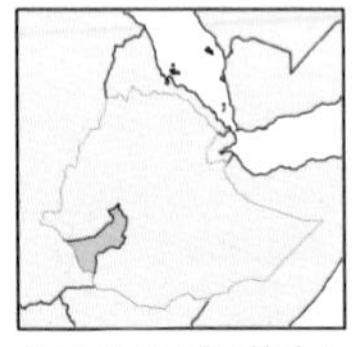

Angraecum infundibulare

The specific epithet '*infundibulare*' refers to the infundibuliform shape of the lip. John Lindley described it in 1862 from a plant collected in Principe by Charles Barter.

Plant pendent. Stems up to 60 × 0.4–0.7 cm. Leaves distichous, narrowly elliptic-ligulate or oblanceolate, unequally obliquely roundly or obtusely bilobed at the

A. infundibulare

apex, 11–21.5 × 1.5–3 cm, twisted at the base and articulated to leaf-sheaths 2.2–5 cm long. Inflorescences 1-flowered; peduncle 3–5.5 cm long. Bracts elliptic-ovate, 6 mm long. Flowers white, tinged with green on the spur, suffused with yellow inside lip, fragrant; pedicel and ovary 5.5–6.5 cm long. Sepals linear-lanceolate, 6–8.5 × 0.3–1 cm. Petals linear, 5.5–6 × 4–5 mm. Lip concave, ovate-elliptic to oblong-ovate, 6.5–8.5 × 5–5.6 cm, with an apicule up to 12 mm long; spur funnel-shaped at the base, 16–21 × 1.5 cm.

Habitat and distribution In former rainforest in very deep valleys, on *Polyscias ferruginea* with tree ferns and *Costus afer* at about 1350 m in Kefa. Also in Nigeria, Cameroon, Uganda, Kenya, and (possibly) Principe.

Flowering period July.

Conservation status Endangered in Ethiopia but not elsewhere in West Africa.

Notes Readily recognised by the large solitary trumpet-shaped flowers with a characteristic bent spur.

2. A. minus *Summerh.*

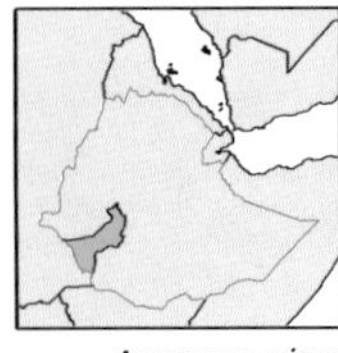
Angraecum minus

The specific epithet '*minus*' refers to the small size of the plants and flowers. Victor Summerhayes described it in 1958 from plants collected in Zimbabwe by Hiram Wild.

A dwarf plant. Stem 2–10 mm long. Leaves distichous, 1.4–5.2 × 1–2 mm. Inflorescences 3–5.2 cm long, 3–14-flowered; peduncle slender, wiry, 1.4–2.1 cm. Bracts broadly ovate, 0.5–1 mm long. Flowers white with a green spur; pedicel and ovary 1 mm long. Dorsal sepal elliptic, 1–1.1 × 0.7 mm; lateral sepals obliquely oblong or obovate, 1.3–1.5 × 0.5–0.6 mm. Petals lanceolate, 1.2 × 0.6–0.7 mm. Lip broadly ovate, 1.4 × 1.6 mm wide; spur conical, 0.6 mm long.

Habitat and distribution The species is found in crowns of trees in forest with *Aningeria altissima*, *Antiaris toxicaria*, *Lannea welwitschii*, *Ficus mucuso*, *Celtis gomphophylla* and *C. zenkeri* at about 1100 m in Kefa. Also in Tanzania, Zambia and Zimbabwe.

Flowering period August and September.

Conservation status Rare and endangered species in Ethiopia but possibly overlooked because of its small size and tiny flowers.

3. A. humile *Summerh.*

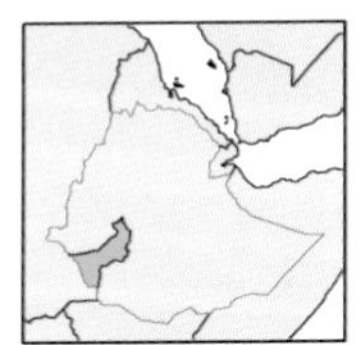
Angraecum minus

The specific epithet '*humilis*' refers to the low-growing habit. Victor Summerhayes described it in 1958 from plants collected in Kenya.

A dwarf plant with a short leafy stem and fine roots. Leaves bilaterally flattened, 6–7 × 2–3 mm, light green. Inflorescences 2–4-flowered, 2–3 mm long; rhachis zigzag. Flowers tiny, greenish white. Sepals elliptic-oblong, 1.4–1.7 × 1.3–1.4 mm. Petals smaller, oblong-elliptic, 1.3–1.4 × 0.7–0.8 mm. Lip concave, ovate, 1.6–1.7 × 1.3–1.5 mm; spur clavate, parallel to ovary, 1.4–1.8 mm long. Column very short, 0.3–0.4 mm long.

Habitat and distribution In tree crowns in semi-natural *Coffea* forest with *Aningeria adolfi-fridericii*, *Phoenix reclinata* and *Dracaena steudneri* at about 1700 m in Kefa. Also in Kenya and Tanzania.

Flowering period Unknown.

Conservation status Rare and endangered in Ethiopia but probably overlooked because of its size and tiny flowers.

29. MICROCOELIA *Lindl.*

Leafless epiphytic or rarely lithophytic herbs with short stems. Roots firmly or loosely attached to the substrate, often dense, terete or less commonly dorso-ventrally flattened, smooth or rarely verrucose, unbranched or with a few branches, usually elongate. Scales on stem protecting the stem apex, acute to rostrate. Inflorescences few–many, axillary, racemose, concentrated in the apical part of the stem, few–many-flowered; peduncle long or short; rhachis terete or angular, smooth or with processes; bracts sheathing or not. Flowers small to minute, more or less sessile or pedicellate, usually white variously tinged with green, brown or pink on spur and other segments. Sepals and petals free, subsimilar. Lip entire or obscurely 3-lobed, free, usually with fleshy calli at the base either side of the mouth of the spur; spur globose, cylindrical or variously swollen. Column fleshy; androclinium short to long; anther-cap hemispherical, often elongated at the apex; pollinia 2, subglobose to pyriform; stipes linear to oblanceolate, entire or bifid at the apex; viscidium linear or oblong, short to long; rostellum bifid, short to as long as the column.

M. globulosa

A genus of 27 species from Madagascar and tropical and southern Africa. A single species is found in Ethiopia.

M. globulosa *(Hochst.) L.Jonss.*

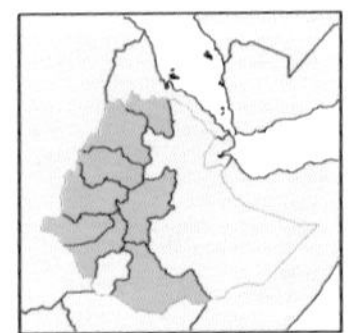

Microcoelia globulosa

The specific epithet '*globulosa*' refers to the globose flowers. Christian Hochstetter described it as *Angraecum globulosum* in 1844 from a plant collected in Tigray by Georg Wilhelm Schimper. Lars Jonsson transferred it to *Microcoelia* in 1981.

A dwarf plant with a short to long stem, up to 65 × 2–3 mm. Scale-leaves acuminate to obtuse, up to 3.5 mm long, with 5–7 nerves. Inflorescences 5–10, stiff, up to 90 mm long, each with up to 15 flowers; peduncle 15 mm long; rhachis flexuose, variously furrowed and excavated with hook-shaped processes below the bracts. Bracts ovate, up to 2.5 mm long. Flowers 5–6 mm long, white, base of perianth and ovary yellow-green, tip of spur orange-brown; pedicel forming a distinct angle with the ovary, up to 3 mm long; ovary 0.3–0.8 mm long. Dorsal sepal ovate, 2.2–3.1 × 1.0–1.4 mm; lateral sepals ovate to narrowly ovate, 2.5–3.5 × 0.9–1.3 mm. Petals obovate to elliptic, 2.1–3 × 0.8–1.2 mm. Lip pandurate, 2.2–3.3 × 1.2–1.8 mm; spur 2.1–3.3 mm long, apex obtuse.

Habitat and distribution On margins of evergreen forest, relict rainforest, riverine forest, and secondary growth between 650 and 2150 m in Tigray, Gonder, Gojam, Shewa, Sidamo, Kefa, Illubabor and Wellega. Also in Uganda, Kenya, Tanzania, Nigeria, south to Angola, Zambia, Malawi and Zimbabwe.

Flowering period January to June.

Conservation status Vulnerable.

M. globulosa

30. DIAPHANANTHE *Schltr.*

Epiphytic or rarely lithophytic herbs with short or long monopodial stems, usually unbranched and covered by sheathing leaf-bases; roots elongate, emerging all along stem through the leaf-bases, often prominent, rarely branched. Leaves distichous, coriaceous or fleshy, rarely thin-textured, unequally bilobed at apex, articulated at base to a sheathing leaf-base, often twisted at base to lie in one plane. Inflorescences 1–many, emerging through the sheathing leaf-bases in upper part of stem, usually several–many-flowered; bracts usually amplexicaul, rarely prominent. Flowers white, pale green or yellow, rarely pinkish, usually translucent, rarely showy. Sepals free, subsimilar. Petals free. Lip entire or obscurely lobed, spurred, usually with a tooth-like or transverse callus in the mouth of the spur. Column porrect; stipites 2; viscidia 1 or 2; rostellum pendent or reflexed, linear or tapering and bifid to pendent, clavate and obscurely 3-lobed.

A genus of about 50 species confined to continental, and mostly tropical, Africa. Six species have been reported from Ethiopia.

Key

1	Stem up to 5 cm long; leaves in a fan	2
–	Stem at least 7 cm long, often much longer; leaves scattered along stem	3
2	Flowers yellow-green; spur 11–12 mm long	**4. D. rohrii**
–	Flowers white; spur 3–4 cm long	**2. D. candida**
3	Leaves slender, 1.5–5 mm wide; flowers minute; sepals and petals 2–3.5 mm long; spur clavate, 1.5–2 mm long	**1. D. adoxa**
–	Leaves broader, more than 1 cm wide; flowers larger; sepals and petals more than 7 mm long; spur more than 5 mm long	4
4	Flowers white; spur slender from a broad mouth, 15–25 mm long	**6. D. tenuicalcar**
–	Flowers yellow, brown or greenish; spur less than 12 mm long	5
5	Leaves oblong to oblanceolate, 6.5–14 cm long; inflorescence less than 8.5 cm long	**5. D. schimperiana**
–	Leaves falcate, oblong to oblanceolate, 10–45 cm long; inflorescence more than 10 cm long	**3. D. fragrantissima**

D. adoxa

1. D. adoxa *F.N.Rasm.*

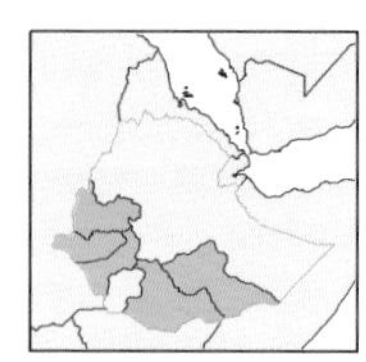
Diaphananthe adoxa

The specific epithet '*adoxa*' refers to the obscure flowers. Finn Rasmussen described it in 1974 from a plant collected in Kefa.

Plant 5–75 cm long. Stems pendent, 2–2.5 mm diameter, often forming dense clumps. Leaves 5–12 in upper part of stem, falcate, linear, unequally bilobed at apex, 35–120 × 1.5–5 mm. Inflorescences 1–several, 2–3.5 cm long, 3–8-flowered; peduncle 2–7 mm long. Bracts amplexicaul, 2 mm long. Flowers green or pale yellowish green; pedicel and ovary up to 2 mm long. Dorsal sepal lanceolate, 2–2.5 × 1 mm; lateral sepals lanceolate, 3–3.2 × 1 mm. Petals ovate, 2.1 × 1.2 mm. Lip entire, ovate, 2 × 2 mm, ecallose; spur clavate, 1.5–2 mm long.

Habitat and distribution In upland evergreen forest and riverine forest between 1300 and 2300 m in Bale, Sidamo, Kefa, Illubabor and Wellega. Also in Uganda and Kenya.

Flowering period July to September.

Conservation status Vulnerable.

Notes This tiny orchid differs from all other Ethiopian species in the genus by its minute flowers with 1.5–2 mm long spurs.

2. D. candida *P.J.Cribb*

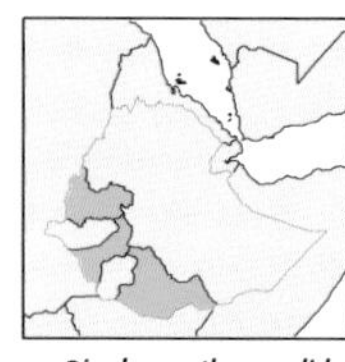
Diaphananthe candida

The specific epithet '*candida*' refers to the pure white flowers. Phillip Cribb described it in 1979 from a plant collected 40 km from Ghimbi on the Asosa road in Wellega by Mike Gilbert and Mats Thulin.

Plant up to 15 cm high. Stem short, leafy, 3 mm in diameter. Leaves 11–16 × 0.9–2 cm, linear, obscurely bilobed at the apex, articulated at the base. Inflorescence 6–14-flowered, 10–22 cm long, peduncle glabrous. Bracts ovate, 3–4 × 3 mm. Flowers pure glossy white, becoming yellowish-brown with age. Dorsal sepal oblong-ovate or lanceolate, 7–8 × 2–3 mm. Lateral sepals lanceolate, 10–11 × 2.5 mm. Petals oblique, ovate, 7 × 3–3.5 mm. Lip deflexed, entire, narrowly-elliptical, 9–10 × 4–5 mm; spur cylindrical, curved, 3.2–3.4 cm long, entrance bearing a fleshy tooth-like callus.

Diaphananthe candida

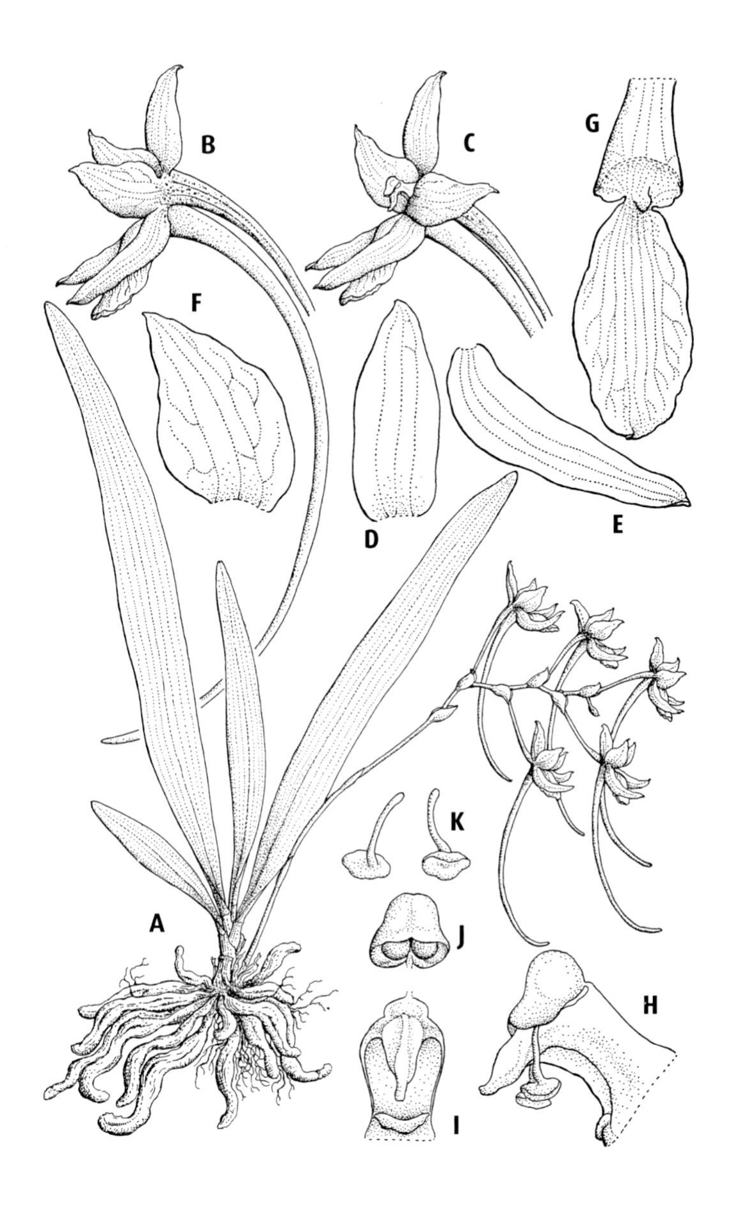

A habit, × ⅔; **B** flower, × 3⅔; **C** flower with petal turned back, × 2; **D** dorsal sepal, × 4; **E** lateral sepal, × 4; **F** petal, × 4; **G** lip, × 4; **H** column, side view, × 10; **I** column, front view, × 10; **J** anther cap, × 10. **K** viscidia and stipites, × 10. All drawn from *Gilbert & Thulin* 802 by Mrs. M. E. Church.

Habitat and distribution An epiphyte on *Acacia* on rocky hillsides at about 2050 m in Sidamo, Kefa and Wellega. Unknown elsewhere.

Flowering period August and September.

Conservation status Endangered.

Notes With its pure white flowers and long spur, *D. candida* looks superficially more like an *Aerangis* than a *Diaphananthe*. However, its lip has a basal tooth in the mouth of the spur and its pollinaria are quite distinct. It differs from *D.rohrii* in its white flowers with a longer spur.

3. D. fragrantissima *(Rchb.f.) Schltr.*

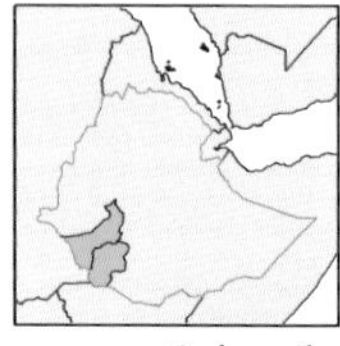

Diaphananthe fragrantissima

The specific epithet '*fragrantissima*' refers to the highly scented flowers. H.G. Reichenbach described it as *Listrostachys fragrantissima* in 1865, based on a collection by Friedrich Welwitsch from Pungo Adongo, Angola.

Plant with long pendulous stems, 7–50 × 5–10 mm. Leaves up to 9, distichous, falcate, linear-oblanceolate, unequally bilobed at apex, 10–44 × 1–3.5 cm, twisted at base. Inflorescences pendulous, 1–several, up to 60 cm long, up to 60-flowered; peduncle 2–20 cm long. Bracts amplexicaul, 3–5 mm long. Flowers usually borne in whorls of up to 4, greenish yellow, diurnally scented; pedicel and ovary 1–2.5 mm long. Dorsal sepal linear, 7–10.5 × 2–2.2 mm; lateral sepals linear, acute, 9.7–11.2 × 2–3 mm. Petals linear to obovate, 8.3–8.5 × 2–2.2 mm. Lip rectangular, 3-lobed at apex, 10–13 × 6.5–7 mm, with erose lateral margins, with a marked tooth at the mouth of the spur; midlobe elongated; spur geniculate at base, 6–9 mm long.

D. fragrantissima

Habitat and distribution In forest, bushland and rocky country, epiphytic on *Acacia* or on rocks between 1100 and 1900 m in Gamo Gofa and Kefa. Also in Burundi, Cameroon, Rwanda, DR Congo, Sudan, Uganda, Kenya, Tanzania, Angola, Malawi, Mozambique, Zambia, Zimbabwe and South Africa.

Flowering period August to November.

Conservation status Endangered.

Notes This very fine species can be readily recognised by its inflorescence being over 10 cm long with flowers borne in whorls. The denticulate lip margin is also distinctive.

4. D. rohrii *(Rchb.f.) Summerh.*

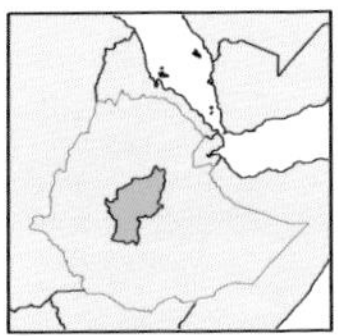
Diaphananthe rohrii

The specific epithet '*rohrii*' was given in honour of Julius von Rohr, the collector of the type specimen. H.G. Reichenbach described it as *Angraecum rohrii* in 1881 from a plant collected in Ethiopia, without exact locality. Victor Summerhayes transferred it to *Diaphananthe* in 1960.

Plant with irregular short stems, 2–5 × 5–7 mm. Leaves 2–7, linear to obovate, slightly unequally bilobed at the apex, 5–15 × 0.7–2.2 cm, articulated to a 5–10 mm long base. Inflorescences 5–17 cm long, laxly many-flowered; penduncle 3–4 cm long; rhachis zig-zag; bracts amplexicaul, 2–8 mm long. Flowers waxy, yellow; pedicel and ovary 4 mm long, scabrid. Dorsal sepal oblanceolate, 3.6–4 × 1.5–2 mm; lateral sepals linear to obovate, 3.7–4 × 1–1.4 mm. Petals linear, 3.4–4.3 × 1 mm. Lip ovate, 3.4–3.8 × 1.7–2.3 mm, with an obscure tooth in the mouth of the spur; spur incurved, clavate, 11–12 mm long.

Habitat and distribution In montane forest between 2100 and 3000 m in Shewa. Also in Liberia, Ivory Coast, Ghana, Togo, Cameroon, São Tomé, Fernando Po, DR Congo, Burundi, Uganda, Tanzania, Kenya and Angola.

Flowering period April to June.

Conservation status Rare and vulnerable.

Notes *D. rohrii* differs from *D. candida* by the flowers being yellow-green rather than white and in having an 11–12 mm rather than 30–40 mm long spur.

5. D. schimperiana *(A.Rich.) Summerh.*

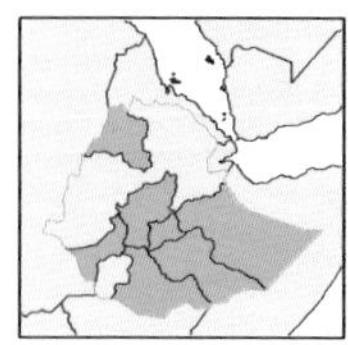

Diaphananthe chimperiana

The specific epithet '*schimperiana*' was given in honour of Georg Wihelm Schimper, the collector of the type specimen. Achille Richard described it as *Dendrobium schimperianum* in 1850 from a plant collected on Mt. Taber (Aber), Gonder. Victor Summerhayes transferred it to *Diaphananthe* in 1945.

Pendent with stems up to 40 cm long, 3–5 mm diameter. Leaves falcate, narrowly oblong to oblanceolate, unequally and roundly bilobed at apex, 6.5–14 × 1.3–2.5 cm, twisted at base to lie in one plane, articulated to a sheathing leaf base 1.4–1.7 cm long. Inflorescences 5.5–8.5 cm long, laxly 5–10-flowered; peduncle 1–1.5 cm long. Bracts obconical, 2–3.5 mm long. Flowers translucent white or cream; pedicel and ovary 2–4 mm long. Dorsal sepal ovate, 5–5.6 × 2.8–3.4 mm; lateral sepals falcate, lanceolate, 7–7.5 × 1.8–2 mm. Petals obliquely triangular-ovate, 5–5.7 × 3–3.6 mm, erose. Lip obscurely 3-lobed in basal half, oblong-ovate in outline, 6–7.7 × 6.6–7.5 mm, with a tooth-like callus in the mouth of the spur; spur slightly dilated in middle, 8–9 mm long.

Habitat and distribution An epiphyte on *Podocarpus* or on rocks in montane forest, between 2100 and 2850 m in Gonder, Shewa, Arsi, Harerge, Bale, Sidamo and Kefa. Also in Sudan and Uganda.

Flowering period April to June.

Conservation status Vulnerable.

Notes *D. schimperiana* differs from *D. tenuicalcar* in its larger flowers and shorter spur.

6. D. tenuicalcar *Summerh.*

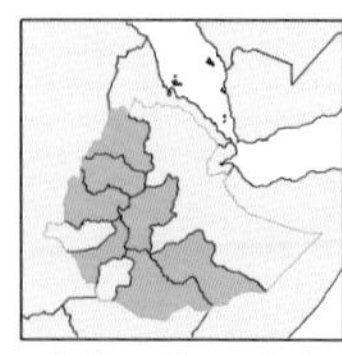

Diaphananthe tenuicalcar

The specific epithet '*tenuicalcar*' refers to the slender spur of the flowers. Victor Summerhayes described it in 1945 from a plant collected by Arthur Stocker Thomas in Karamoja District, Uganda.

Plant often growing in masses, with spreading or pendulous stems, 9–26 × 1–3 mm. Leaves 3–8, towards stem apex, falcate, linear-lanceolate, obliquely bilobed at acute apex, 10–60 × 7–13 mm, twisted at base and articulated. Inflorescences 1–many, shorter than or equal to leaf, 1–3 cm long, up to 7-flowered; peduncle 5–7 mm long. Bracts amplexicaul, conical, 2 mm long.

Flowers white, sweetly scented; pedicel and ovary 4–12 mm long. Dorsal sepal oblong-elliptic, 4–4.5 × 2–2.8 mm; lateral sepals obliquely ovate or elliptic, 4.2–5.1 × 2–2.5 mm. Petals obliquely ovate, 4.2–4.8 × 2.5 mm. Lip flabellate, obscurely 2–4-lobed on apical margin, 4.5–6.5 × 6.2–7.6 mm, with a tooth in the mouth of the spur; spur slender, cylindrical from a broad mouth, 15–25 mm long.

Habitat and distribution Found in and at the edge of montane forest and in wooded grassland between 1350 and 2800 m in Gonder, Gojam, Shewa, Bale, Sidamo, Kefa and Wellega. Also in Uganda and Kenya.

Flowering period June to September; December.

Conservation status Vulnerable.

Notes *D. tenuicalcar* differs from *D. schimperiana* by its smaller flowers and much longer spur.

D. tenuicalcar

31. BOLUSIELLA *Schltr.*

Erect epiphytic or occasionally lithophytic, entirely glabrous monopodial herbs. Stems abbreviated, simple, leafy above, with numerous tufted roots below. Leaves equitant, imbricate, arranged in a vertical plane forming a fan, narrowly elliptic or ensiform to linear-ligulate, obtuse to acute, rigid and articulated to a sheathing base, fleshy, sometimes sulcate-canaliculate and falcate-recurved. Inflorescence exceeding leaves, densely or laxly many-flowered, arising from axils of old leaf-sheaths; peduncle shorter than or equalling rhachis, rarely longer; bracts small or large, imbricate and partially concealing flowers, membranous, grey to blackish brown or olive. Flowers resupinate, very small, white. Tepals free, oblong or narrowly elliptic, obtuse to acuminate, subequal, spreading. Lip entire or obscurely 3-lobed, spurred, oblong or ovate-elliptic, obtuse to acuminate, sometimes recurved; spur cylindrical, ellipsoid or conical, obtuse, usually shorter than lip. Column oblong, fleshy, constricted above, without a foot; rostellum subulate, hooked; anther quadrate, cucullate; pollinia 2, oblong or ellipsoid, each attached by a linear stipe to a large ovate viscidium.

A genus of five species distributed in tropical and southern Africa. A single species is found in Ethiopia.

B. iridifolia (*Rolfe*) *Schltr.*

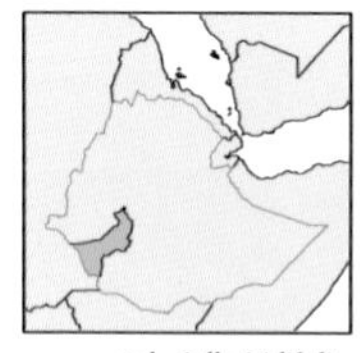

Bolusiella iridifolia subsp. iridifolia

The specific epithet '*iridifolia*' refers to the iris-like leaves. Robert Rolfe described it as *Listrostachys iridifolius* in 1897, based on a collection by Friedrich Welwitsch from Golungo Alto, Angola. Rudolf Schlechter transferred it to *Bolusiella* in 1918.

subsp. **iridifolia**

Dwarf herb. Stem short, 0.5–2 cm long. Leaves 4–10, ensiform, 10–45 × 1.5–4 mm, sulcate on upper surface. Inflorescence dense, 3.5–6 cm long. Bracts ovate, brown or black, shorter than or longer than flowers, 1–4 mm long. Flowers white; pedicel with ovary 2 mm long. Tepals oblong, 2–3 × 0.7–1.4 mm. Lip oblong-ligulate, slightly sigmoid, 2.5–5 × 0.7–1.1 mm; spur cylindrical-ellipsoid and inflated or conical and saccate, obtuse, strongly incurved at right-angles to lip, 1.5–2 mm long.

Habitat and distribution An epiphyte in canopy of *Aningeria altissima* and *Antiaris toxicaria* forest or rarely on rocks between 950 and 1100 m in Kefa. Also in Ivory Coast, Ghana, Equatorial Guinea, Cameroon, Uganda, Kenya, Tanzania, DR Congo, Angola and Grand Comore.

Flowering period September.

Conservation status Endangered but possibly overlooked because of its diminutive stature.

Notes Readily distinguished from all other Ethiopian orchids by its dwarf stature, fan-like habit and tiny flowers. *B. iridifolia* subsp. *picea* from tropical East Africa has a shorter, conical-saccate, more or less straight spur.

32. AERANGIS *Rchb.f.*

Epiphytic herbs with short or elongated woody stems bearing numerous elongated aerial roots in the lower part. Stems bearing few–several leaves apically. Leaves in 2 rows, much longer than broad and usually wider in the upper half, unequally bilobed at the apex. Inflorescence lateral, a short or elongated raceme, rarely branched, few–many-flowered. Flowers resupinate, white or variously tinted with green or brown. Sepals and petals free, spreading or reflexed. Lip entire, often similar to the sepals and petals, spurred at the base. Column short and stout or somewhat elongated and more slender, often narrowed towards the base and enlarged at the level of the stigma; androclinium straight or sloping, the anther-cap sometimes beaked; rostellum entire, elongated, deflexed or porrect; pollinia 2, sessile on a single stipe; viscidium variously shaped; stigma an oval or rhomboid sticky depression. Ovary elongate, straight or curved. Capsule cylindric or ellipsoid, often much elongated.

A genus of approximately 50 species of which at least 26 occur in Africa and the remainder in Madagascar and the Comoro Islands. One species is recorded from both East Africa and Sri Lanka, where it is probably introduced. Several of the more widely distributed species have been known in cultivation for many years. Five species have been reported from Ethiopia.

A. luteo-alba var. *rhodosticta*

Key

1	Spur less than twice as long as lip; column bright red	**1. A. luteo-alba** var. **rhodosticta**
–	Spur much longer than the lip, usually more than 3 times as long; column white	2
2	Spur loosely spiralling	3
–	Spur straight	4
3	Stem 20–100 cm long; spur 10–15 cm; lip elliptic-lanceolate, 20–25 mm long, 7–8 mm broad in basal half when flattened	**2. A. thomsonii**
–	Stem usually less than 10 cm long; spur 13–22 cm; lip subpandurate, 15–20 mm long, 8–15 mm broad in apical half	**3. A. kotschyana**
4	Leaves grey-green; sepals and petals oblong-elliptic, 9–14 mm long	**4. A. somalensis**
–	Leaves dark green; sepals and petals lanceolate, acuminate, 20–45 mm long	**5. A. brachycarpa**

1. A. luteo-alba (*Kraenzl.*) *Schltr.*

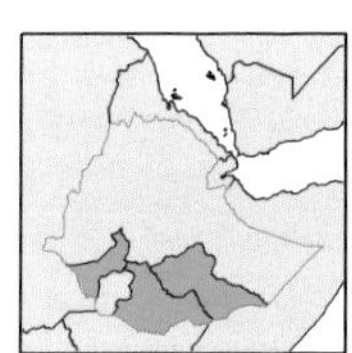

Aerangis luteoalba var. *rhodosticta*

The specific epithet '*luteoalba*' refers to the yellowish white flowers of the type. Fritz Kraenzlin described it as *Angraecum luteoalbum* in 1895 from plants collected by Franz Stuhlmann in DR Congo. Rudolf Schlechter transferred it to *Aerangis* in 1918.

var. rhodosticta (*Kraenzl.*) *J.Stewart*

Fritz Kraenzlin described it as *Angraecum rhodostictum* in 1896, based on collections by Count E. Ruspoli and Domenico Riva in Ethiopia and Georg August Zenker and Alios Staudt in Cameroon. Joyce Stewart considered it a variety of *Aerangis luteo-alba* in 1979.

Stem 1–3 cm long, often pendent with numerous thin flexuous roots which are relatively long for the size of the plant. Leaves 2–8, lying in one plane, linear-ligulate or linear-lanceolate, up to 15 × 6–15 mm, sometimes falcate, unequally bilobed at the apex, the lobes rounded. Inflorescences arising from the stem below the leaves, arching or pendent racemes, few–25-flowered in the upper half or two-thirds, up to 35 cm long; peduncle slender, terete; rhachis similar, slightly zigzag. Bracts spreading or slightly reflexed, ovate, 3–4 mm long. Flowers arranged in 2 rows, all in the same plane, 7–20 mm apart; perianth parts all spreading, white, cream, greenish, or yellowish white; column similar in colour, or bright red; pedicel and ovary slender, 1–2 cm long. Sepals oblanceolate, 10–15 × 37 mm. Petals obovate or broadly oblanceolate, 5–10 mm. Lip obovate or rhombic, widest above the middle, 5–20 × 7–15 mm; spur slightly thickened in the apical half, 2.3–4 cm long.

Habitat and distribution In montane forest and on coffee bushes near swampy areas between 1500 and 2200 m in Bale, Sidamo and Kefa. Also in Central African Republic, Cameroon, DR Congo, Uganda, Kenya and Tanzania.

Flowering period August to November.

Conservation status Locally common but vulnerable in Ethiopia. Locally common elsewhere but often collected for its horticultural merit.

Notes This distinctive species has broad sepals and petals and a bright red rather than white column in the centre of the flower. The typical variety from DR Congo differs from var. *rhodosticta* in having a creamy white column.

2. A. thomsonii *(Rolfe) Schltr.*

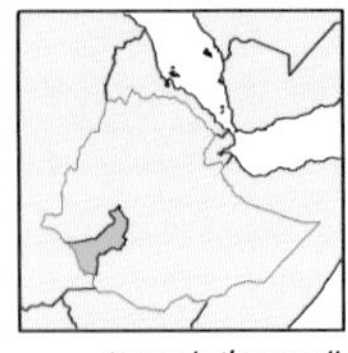

Aerangis thomsonii

The specific epithet '*thomsonii*' was given in honour of Joseph Thomson, who collected the type specimen in Kenya. Robert Rolfe described it as *Angraecum thomsonii* in 1897. Rudolf Schlechter transferred it to *Aerangis* in 1918.

Stem woody, up to 10 × 10–100 cm long. Leaves 8–20, alternate, distichous, ligulate, 8–28 × 1.5–4.5 cm, apex unequally lobed. Inflorescences borne at the nodes, arching racemes, 4–10-flowered; peduncle up to 9 × 6 mm; rhachis more or less straight, green, up to 20 cm long; bracts very broadly triangular-ovate, cucullate, 10–15 mm long. Flowers held erect in 2 rows, 1.5–3.5 cm apart, white; pedicel and ovary green, 3–6 cm long. Dorsal sepal lanceolate-elliptic, 22–30 × 7–9 mm; lateral sepals lanceolate-elliptic, 25–32 × 5–6 mm. Petals lanceolate-elliptic, 20–25 × 6–8 mm. Lip elliptic-lanceolate, 20–25 × 7–8 mm; spur cylindrical but widened and flattened in its terminal half, 10–15 cm long.

Habitat and distribution In shady places, usually rather low down on trunks and branches in highland forest between 1600 and 2600 m in Kefa. Also in Uganda, Kenya and Tanzania.

Flowering period April to June.

Conservation status Endangered.

Notes *A. thomsonii* is readily distinguished from all other Ethiopian *Aerangis* by its long stem and large flowers.

3. A. kotschyana *(Rchb.f.) Schltr.*

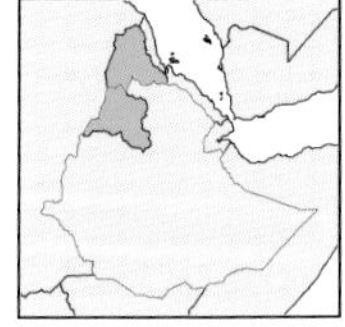

Aerangis kotschyana

The specific epithet '*kotschyana*' was given in honour of Theodore Kotschy, who collected the type specimen in Gonder. H.G. Reichenbach described it as *Angraecum kotschyanum* in 1864. Rudolf Schlechter transferred it to *Aerangis* in 1918.

Stem stout and woody, upright, up to 20 × 1.5 cm. Leaves 3–20, distichous, obovate-oblong, 6–30 × 2–8 cm, unequally or subequally bilobed at the apex. Inflorescences axillary, often several, arching or pendent racemes, up to 20-flowered; peduncle terete, up to 12 cm long. Bracts triangular, ovate, up to 8 mm long. Flowers in 2 rows, white, often tinged with salmon pink at the centre and in the tips of the perianth; pedicellate ovary pinkish green, 2–3 cm long. Dorsal sepal elliptic-lanceolate, 12–25 × 9 mm; lateral sepals lanceolate, 14–28 × 7 mm. Petals oblanceolate, 10–23 × 8 mm. Lip 10–20 × 8–15 mm; spur pendulous, thickened in its lower third and curved or twisted just below the middle, 13–22 cm long.

A. kotschyana

Habitat and distribution An epiphyte on old trees in humid wooded grassland, woodland, and sometimes in forests between sea-level and 1500 m in Gonder and in Eritrea. Also in Guinea, Nigeria, Burundi, Central African Republic, Rwanda, DR Congo, Sudan, Kenya, Tanzania, Uganda, Zanzibar, Malawi, Mozambique, Zambia and Zimbabwe.

Flowering period May to July.

Conservation status Rare and vulnerable.

Notes *A. kotschyana* differs from *A. thomsonii* by its shorter stem and flowers with a long spiralling spur.

4. A. somalensis (*Schltr.*) *Schltr.*

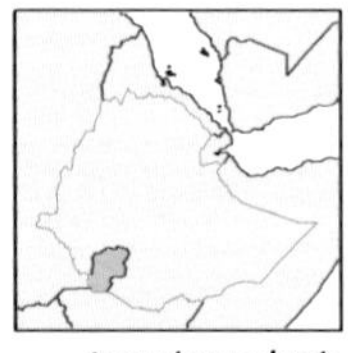

Aerangis somalensis

The specific epithet '*somalensis*' refers to the Ogaden region in Eastern Ethiopia, which is inhabited by Somali ethnic groups, from where it was collected by the Italian explorers Count E. Ruspoli and Domenico Riva. Rudolf Schlechter described it as *Angraecum somalense* in 1906. He transferred it to *Aerangis* in 1918.

Stem short, upright, bearing numerous greyish, flexuous roots, 5–7 mm in diameter. Leaves 2–6, distichous, oblong-ligulate, 2–11 × 1.3–3.4 cm unequally or subequally bilobed at the apex. Inflorescences arising below the leaves, usually several, 10–20 cm long, 4–17-flowered. Bracts brownish, ovate, up to 5 mm long. Flowers white, sometimes tinged with pink, 1–2 cm apart; pedicellate ovary 1.5–2 cm long. Dorsal sepal ovate, 8–10 × 4–6 mm; lateral sepals oblong-ligulate, 9–14 × 3–5 mm. Petals oblong-apiculate, 8–11 × 3–5.5 mm. Lip oblong-ligulate, the margins reflexed in the lower half, 9–13 × 4–7 mm; spur sometimes slightly inflated in the lower half but narrowing again towards the tip, 10–15 cm long. Column rather thick, to 4 mm high; anther-cap slightly beaked; pollinia globose, 0.8 mm long; stipes slender but enlarged in the upper third; viscidium ovoid, 1–1.5 long with outer margins rolled over the pointed rostellum.

Habitat and distribution On trees and at the base of branching shrubs in relict patches of woodland along streams and near rock outcrops in dry areas between 1000 and 1200 m in Gamo Gofa. Also in Kenya, Tanzania, Zimbabwe and South Africa (Transvaal).

Flowering period April.

Conservation status Endangered and possibly extinct because of habitat destruction.

Notes Readily recognised by its short broad sepals and petals and bluish green leathery leaves in which the veins are clearly marked.

A. somalensis

5. A. brachycarpa *(A.Rich.) Th.Dur.& Schinz*

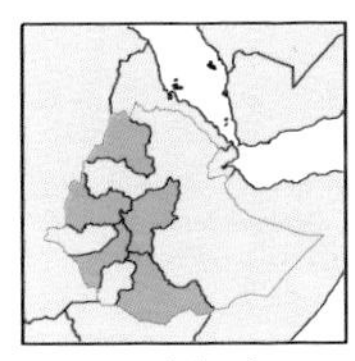

Aerangis brachycarpa

The specific epithet '*brachycarpa*' refers to the relatively short fruit. Achille Richard described it as *Dendrobium ? brachycarpum* in 1850 from a plant collected by Georg Wilhelm Schimper in Wogera, Gonder. Théophile Durand and Hans Schinz transferred it to *Aerangis* in 1892.

Stem woody, upright or pendent, up to 20 cm long but usually much shorter. Leaves 4–12, alternate, usually distichous, obovate or spathulate, up to 25 × 2–6 cm, apex unequally bilobed. Inflorescences axillary racemes, arching or pendent, up to 40 cm long, 2–12-flowered; peduncle terete up to 10 × 3 mm; rhachis slightly zigzag, gradually tapering, greyish green, black-dotted; bracts triangular, up to 9 mm long. Flowers in 2 rows, pale green when first opening, becoming white, and often with a pink or brownish tinge in the petals and spur; pedicel and ovary straight or arching, 3.5–7 cm long. Tepals all narrowly lanceolate, 20–45 × 4–8 mm; dorsal sepal erect, lateral sepals and petals becoming sharply reflexed within a few days of the flowers opening. Lip deflexed, 5–10 mm wide near the base. Spur gradually tapering towards the tip, which is sometimes minutely bifid, 12–20 cm long.

Habitat and distribution In dense shade, usually rather low down on tree trunks, branches, or in the forking bases of bushes in highland forests and grassland with scattered trees between 1450 and 2300 m in Gonder, Shewa, Sidamo, Kefa and Wellega. Also in Uganda, Kenya, Tanzania, Malawi, Zambia and Angola.

Flowering period March to July.

Conservation status Rare and vulnerable.

Notes Readily distinguished from *A. somalensis* by its thinner green rather than blue-green leaves and more stellate flowers.

A. brachycarpa

33. RANGAERIS *Summerh.*

Epiphytic or rarely lithophytic herbs with short to long stem entirely covered with persistent leaf-bases; roots emerging along the stem opposite the leaves, rather stout, unbranched or little-branching. Leaves coriaceous, distichous, conduplicate, linear or oblong, unequally bilobed at the apex. Inflorescences pendent or spreading to suberect, usually laxly few–several-flowered. Flowers small to relatively large, usually white or yellowish; pedicel and ovary longer than the bract. Sepals and petals free, subsimilar, spreading to recurved. Lip entire or obscurely 3-lobed, lacking a callus; spur pendent, filiform, elongate from a narrow mouth. Column short to long, glabrous or puberulent, lacking a foot; rostellum bifid; anther cucullate, produced in front and truncate at the apex; pollinia 2, pyriform; stipites 2, linear or oblanceolate; viscidium relatively large, oblong or cordate.

A small genus of about six species confined to tropical Africa. A single species is found in Ethiopia.

R. amaniensis *(Kraenzl.) Summerh.*

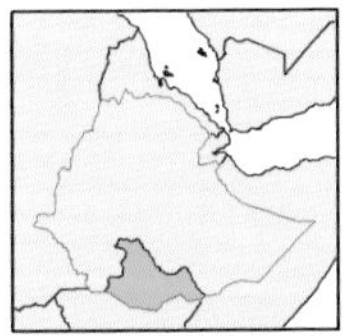
Rangaeris amaniensis

The specific epithet '*amaniensis*' refers to Amani in the Usambara Mountains, Tanzania, from where the type specimen was collected by Karl Braun. Fritz Kraenzlin described it as *Listrostachys amaniensis* in 1909. Victor Summerhayes transferred it to his new genus *Rangaeris* in 1949.

Erect to pendent plant, often forming dense clumps. Stems up to 45 cm long. Leaves coriaceous, conduplicate, narrowly oblong, unequally roundly bilobed at the apex, 3.5–11.5 × 1–2.3 cm. Inflorescences spreading, 6–9 cm long, 5–13-flowered; peduncle 1–1.5 cm long. Bracts 5–6.5 mm long, drying black. Flowers white fading to yellow, greenish on the outer surface and spur, diurnally scented; pedicel and ovary 2.5–5.5 cm long, scabrid. Sepals lanceolate, 10–25 × 2.7–5.6 mm. Petals lanceolate, 9–21 × 2.2–5 mm. Lip obscurely 3-lobed in the middle, 10–24 × 5–11 mm; side lobes rounded; midlobe lanceolate; spur pendent, filiform, 7.5–16 cm long.

Habitat and distribution In montane forest of *Juniperus procera* and in open woodland between 1100 and 2600 m in Sidamo. Also in Uganda, Kenya, Tanzania and Zimbabwe.

R. amaniensis

Flowering period September to November.

Conservation status Very local in Ethiopia. Locally common elsewhere in its range.

34. CYRTORCHIS *Schltr.*

Epiphytic or rarely lithophytic herbs with short to long, erect or pendent stems, covered by distichous leaf-bases; roots emerging all along stem, elongate, branching. Leaves distichous, coriaceous or fleshy, flat or conduplicate, much longer than broad, unequally bilobed at apex, articulated to a sheathing base, deciduous, leaving obvious sharp edges of leaf-bases. Inflorescences 1-several, usually shorter than the leaves, few–many-flowered, axillary in upper part of stem; peduncle short, often covered by sterile sheathing bracts; bracts usually large. Flowers stellate, sweetly scented, white, sometimes with a greenish, pinkish or brownish spur. Sepals and petals subsimilar, free, linear-lanceolate, acuminate, usually recurved. Lip similar, entire or very obscurely 3-lobed, lacking a callus, with a long tapering spur at the base; spur sigmoid or slightly incurved. Column short, fleshy; rostellum 3-lobed, pendent, outer lobes much longer than the midlobe, often papillate towards apex; pollinia 2; stipites 2, oblanceolate; viscidium either hyaline or comprising an indurate saddle-shaped upper part and a hyaline lower part; anther-cap elongated at apex. Fruit a 3–6-winged capsule.

A genus of about 16 species confined to tropical and southern Africa. Two species have been recorded from Ethiopia.

Key

1	Largest leaves up to 11.5 cm long; bracts c. 12 mm long	**1. C. arcuata**
-	Largest leaves 13–15 cm long; bracts less than 10 mm long	**2. C. erythraeae**

1. C. arcuata *(Lindl.) Schltr.*

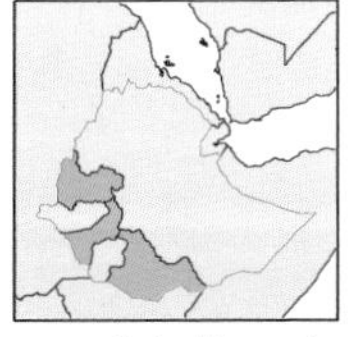
Cyrtorchis arcuata

The Latin epithet '*arcuata*' refers to the curved leaves. John Lindley described it as *Angraecum arcuatum* in 1837 from a plant collected by Johann Drège in Albany, Cape Province, South Africa. Rudolf Schlechter transferred it to *Cyrtorchis* in 1914.

C. arcuata

An epiphytic or lithophytic herb, often forming large masses. Stems at least 15–30 cm long. Leaves linear, narrowly oblong or narrowly elliptic, rarely oblanceolate, 8–11.5 × 1.5–2 cm. Inflorescences 6–20 cm long, 5–14-flowered; peduncle 1.5–2 cm long. Bracts cymbiform, ovate, 15–33 × 5–17 mm. Flowers white with a greenish spur, sweetly scented; pedicel and ovary 20–40 mm long. Sepals lanceolate, 18–49 × 5–7 mm. Petals similar, 14–27 × 4.5–6 mm. Lip lanceolate, 15–36 × 4–6 mm; spur incurved to somewhat sigmoid, 30–70 mm long.

Habitat and distribution On trees and rocks in bush, woodland and forest between 1350 and 2000 m in Sidamo, Kefa and Wellega. Also in Sierra Leone to DR Congo, Uganda, Tanzania, Kenya, Zanzibar and south to Zimbabwe and South Africa.

Flowering period May to August.

Conservation status Vulnerable.

Notes *C. arcuata* is a common, widespread and variable species throughout tropical and South Africa. Three varieties have been recognised but their status is still the subject of debate. It differs from *C. erythraeae* by its larger bracts.

2. C. erythraeae *(Rolfe) Schltr.*

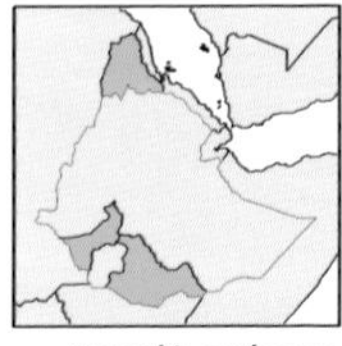
Cyrtorchis erythraeae

The specific epithet '*erythraeae*' refers to Eritrea, from whence the type specimen was collected. Robert Rolfe described it as *Angraecum erythraeae* in 1897 from a plant collected in Eritrea by Georg Schweinfurth. Rudolf Schlechter transferred it to *Cyrtorchis* in 1918.

An epiphytic herb with an erect leafy stem, up to 22 × 1 cm. Leaves distichous, linear, unequally roundly bilobed at apex, 6–11 × 0.7–1.4 cm, twisted near base. Inflorescence 1-several, 2–5.5 cm long, shorter than the leaves, 3–5-flowered. Bracts ovate, cucullate, 9–15 × 5–9 mm, drying brown. Flowers white, stellate; pedicel and ovary 11–24 mm long. Sepals and petals subsimilar, linear-tapering, 12–16 × 2–3 mm; petals slightly falcate. Lip similar to sepals and petals, cucullate, 12–14 × 3–4 mm; spur pendent, incurved, 26–32 mm long.

C. erythraeae

Habitat and distribution An epiphyte on *Olea chrysophylla* in dry scrub forest and secondary forest between 1350 and 1700 m in Sidamo and Kefa and in Eritrea. Unknown elsewhere.

Flowering period May to August.

Conservation status Rare.

Notes *C. erythreae* differs from *C. arcuata* by the smaller bracts. However, its distinctiveness is questionable and further work is necessary to establish its taxonomic status.

35. ANCISTRORHYNCHUS *Finet*

Small to medium-sized monopodial epiphytic herbs. Stems short, thick, covered by persistent sheathing leaf-bases; roots elongate, clustered towards the base of the stem. Leaves imbricate, suberect, spreading or recurved, ligulate or tapering, apex more or less unequally bilobed, lobes sometimes toothed, coriaceous or fleshy. Inflorescences axillary from the lower leaves, almost sessile, usually forming globose or ellipsoidal heads; bracts papery to membranous, as long as the flowers. Flowers mostly white, often marked with green on the lip. Sepals and petals free, subsimilar.

Lip oblong to orbicular, entire to 3-lobed, spurred at the base, lacking a callus; spur straight to sigmoid or geniculate, sometimes swollen at the apex and/or the mouth. Column short, fleshy; rostellum projecting down then abruptly geniculate in the middle and upcurved in the apical half, 2-lobed, each lobe falcate and acute; anther-cap hemispherical, produced in front into a short truncate appendage; pollinia 2, globose; stipites either 2 and subspathulate-oblanceolate or 1 and bifid; viscidium 1, long and narrow.

A genus of about 16 species, widespread in tropical Africa. It is represented by a single species in Ethiopia.

A. metteniae *(Kraenzl.) Summerh.*

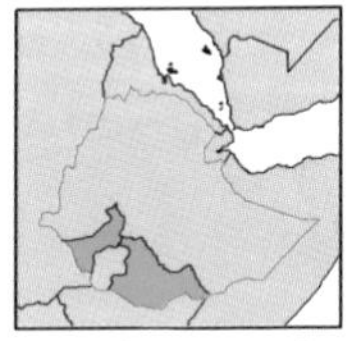

Ancistrorhyncus metteniae

The specific epithet was given in honour of Herr Metten. Fritz Kraenzlin described it as *Listrostachys metteniae* in 1893 from a plant collected in Cameroon by Karl Braun. Victor Summerhayes transferred it to *Ancistrorhynchus* in 1944.

Pendent with a short stem 2–7 cm long. Leaves linear, unequally bilobed at the apex, 16–25 × 1.4–1.6 cm wide. Inflorescences densely many-flowered, 1.5–1.7 cm long. Bracts ovate, 6–10 × 3–5 mm. Flowers white with a green mark on the lip; pedicel and ovary 4–5 mm long, scabrid. Dorsal sepal elliptic, 3.6–5.1 × 2.1–2.3 mm; lateral sepals obliquely oblong, 4–5.1 × 2 mm. Petals elliptic-oblong, 3.7–4.7 × 1.7–2.4 mm. Lip broadly ovate-subcircular, 3.5–4 × 4.3–4.5 mm; spur clavate, straight, 3–4.3 mm long.

Habitat and distribution In evergreen forest of *Aningeria altissima* and light riverine forest between 900 and 1300 m in Kefa and Sidamo. Also in Sierra Leone to Nigeria and the Central African Republic, Uganda and Tanzania.

Flowering period July to September.

Conservation status Rare and possibly endangered in Ethiopia. Locally frequent elsewhere in its range.

A. holochila

36. ANGRAECOPSIS *Kraenzl.*

Small epiphytes or rarely lithophytes with short or very short monopodial stems. Roots long, flexuous, terete or flattened, simple or branching. Leaves distichous, ligulate to oblanceolate, often falcate, unequally bilobed at the apex, sometimes twisted at the base and articulated to imbricate sheathing bases. Inflorescences axilary, 1–several, racemose, few–many-flowered; peduncle wiry, often longer than the rhachis; bracts small. Flowers small, white, pale green or yellow-green. Dorsal and lateral sepals similar or dissimilar; lateral sepals often oblanceolate and produced at one side at the base. Petals often adnate at the base to the lateral sepals and obliquely triangular. Lip 3-lobed or rarely entire, lacking a callus, spurred at the base; spur sometimes inflated at the apex. Column fleshy, short; rostellum short or rarely elongate, 3-lobed; pollinia 2, globose; stipites 2, linear or rarely oblanceolate; viscidia 2 or rarely 1, elongate.

A small genus of about 15 species in tropical Africa, Madagascar and the Mascarene Islands. Two species are found in Ethiopia.

Key

1	Stem less than 5 mm long; inflorescence 4–5 cm long, laxly few–8-flowered; flowers pale green, yellowing with age; lip entire; spur 14–23 mm long, not swollen at apex	**1. A. holochila**
–	Stem 3–6 cm long; inflorescence 10–20 cm long, many flowered; flowers white; lip strongly 3-lobed; spur 10–11 mm long, swollen at apex	**2. A. trifurca**

1. A. holochila *Summerh.*

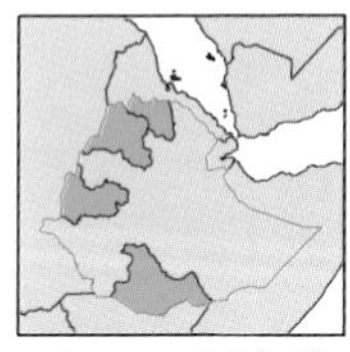
Angraecopsis holochila

The specific epithet '*holochila*' refers to the entire lip. Victor Summerhayes described it from plants collected by Arthur Stocker Thomas in Napak, Karamoja District, Uganda.

A dwarf plant with a stem less than 5 mm long. Leaves linear, 4–9 × 0.4–0.6 cm, twisted and articulated at the base. Inflorescences 4–5 cm long, laxly few–8-flowered; peduncle 1–2 cm long. Bracts ovate, 1–1.5 mm long. Flowers pale green turning yellow with age; pedicel and ovary 5–10 mm long. Dorsal sepal elliptic, 3–4.2 × 1.5–2.9 mm; lateral sepals oblanceolate, 4.5–5.3 × 1.6–2.3 mm. Petals obliquely ovate, 3–3.5 × 1.8–2.7 mm. Lip entire, ovate, 3 × 1.6–1.8 mm; spur narrowly cylindrical, 14–22.5 mm long.

A. *holochila*

Habitat and distribution On trees at edge of montane forest in ravines and on rocks in the open between 1500 and 2300 m in Tigray, Gonder, Sidamo and Wellega. Also in Uganda.

Flowering period July to September.

Conservation status Vulnerable throughout its range.

Notes Differs from *A. trifurca* in its far smaller habit and flowers with an entire lip.

2. A. trifurca *(Rchb.f.) Schltr.*

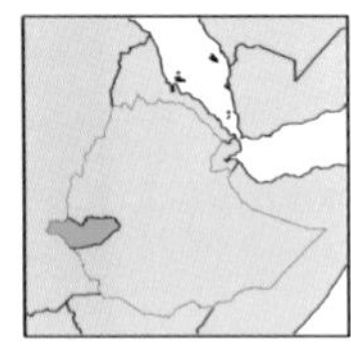

Angraecopsis trifurca

The specific epithet '*trifurca*' refers to the 3-lobed lip. H.G. Reichenbach described it as *Aeranthus trifurcus* in 1885, from a plant collected in the Comores. Rudolf Schlechter transferred it to *Angraecopsis* in 1915.

A pendent epiphyte with a short stem 3–6 cm long. Leaves 2–5, falcate, linear to oblanceolate, unequally acutely or subacutely bilobed at the apex, up to 17 × 3.5 cm. Inflorescences breaking through the old leaf bases, 10–22 cm long, 4–8 flowered; peduncle 5–12 cm long. Flowers whitish-green; pedicel and ovary 9–16 mm long. Dorsal sepal ovate, very convex, 3–3.6 × 3–3.5 mm; lateral sepals unequally obovate, 6–8 × 2–3 mm. Petals obliquely triangular, 2.8 × 3.5 mm, recurved, apex acuminate. Lip distinctly 3-lobed, 4.2–6.5 mm long; side-lobes linear, 2–3 × 0.4–0.6 mm, mid-lobe narrowly triangular, 2.8–3.9 × 0.5–1.0 mm; spur straight, tubular, parallel with pedicel and ovary but a little longer, 14–17 × 0.5–0.8 mm, apical part slightly inflated. Column 2 mm long.

Habitat and distribution In forest on the bole of a tree on the bank of a stream at about 1000 and 1300 m in Illubabor. Also in the Comores.

Flowering period Not observed flowering in Ethiopia; live specimens flowered in October and November in Copenhagen.

Conservation status Endangered in Ethiopia and throughout its range.

Notes Differs from *A. holochila* in its far larger leaves and flowers with a 3-lobed lip. The similar Zambesian species *A. gassneri* has distinctly verrucose stipites, an apically triangular anther cap and corrugated leaves, while *A. macrophylla* (endemic in Uganda) is easily distinguished by the narrower leaves, much more deeply trifurcate lip and straight spur. The widespread *A. parviflora* has generally narrower leaves and distinctly smaller flowers. The Ethiopian material can only tentatively be assigned to *A. trifurca* and may eventually prove to belong to a distinct species, but the scarce material available until now does not allow a clear distinction from *A. trifurca*.

A. trifurca

37. TRIDACTYLE *Schltr.*

Epiphytic or rarely lithophytic monopodial herbs. Stems short to long, unbranched or little branched, erect to pendent stems, covered by sheathing leaf-bases. Leaves thin-textured or fleshy, all along stem, distichous, usually twisted at base to lie in one plane, linear to elliptic or narrowly lanceolate, unequally bilobed at apex, articulated to a sheathing leaf-base. Inflorescences emerging through the leaf-sheaths opposite the leaves, 2–many-flowered; peduncle usually short; rhachis often zigzag; bracts amplexicaul. Flowers usually rather small and seldom showy, white, yellow, orange or green, usually scented; pedicel and ovary short, often lepidote. Sepals and petals subsimilar, free, spreading. Lip entire to 3-lobed, auriculate at base, with a short to long spur at the base. Column short, fleshy, lacking a foot; anther-cap thin-textured, smooth or papillate; pollinia 2; stipes 1, entire or Y-shaped; viscidium circular or elliptic, usually rather small; rostellum elongate, tapering apex, slender.

A genus of about 42 species in tropical Africa and in South Africa. Two species are found in Ethiopia.

Key

1 Leaves dorsiventrally flattened or channelled, 5–15 mm broad; inflorescences 3.5–13 cm long; lip side lobes longer than the midlobe — **1. T. bicaudata**

– Leaves terete, filiform, up to 1 mm broad; inflorescences up to 8 mm long; lip side lobes shorter than the mid-lobe — **2. T. filifolia**

1. T. bicaudata *(Lindl.) Schltr.*

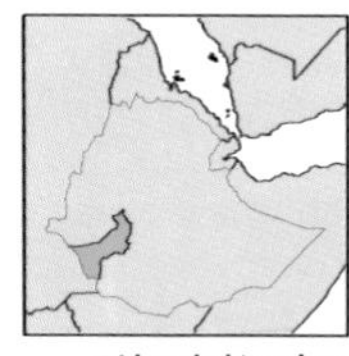

Tridactyle bicaudata

The specific epithet '*bicaudata*' refers to the two tail-like side lobes of the lip. John Lindley described it as *Angraecum bicaudatum* in 1837 from a plant collected by Johann Drège in Zwartkopsrivier, South Africa. Rudolf Schlechter transferred it to *Tridactyle* in 1914.

A pendent plant often forming large hanging masses on trees; stems 12–80 cm × 3.5–6 mm. Leaves linear, unequally roundly bilobed, 9–14.5 × 0.5–1.3 cm, usually twisted at base to lie in one plane or less commonly conduplicate. Inflorescences suberect-spreading, 3.5–13 cm long, 8–25-flowered. Bracts amplexicaul, 1–2 mm long. Flowers white, yellow or

T. bicaudata

greenish yellow, fragrant; pedicel and ovary 2–3 mm long, glabrous. Dorsal sepal oblong-ovate, 4.4–5.9 × 2.3–3.2 mm. Lateral sepals obliquely ovate, 4.7–6 × 2.6–3.6 mm. Petals oblong, acute or obtuse, 3.7–6.2 × 1.2–12.2 mm. Lip 3-lobed in the middle, auriculate at the base, 3–5.7 × 8–11.5 mm; side lobes spreading, linear, 3.6–5.5 mm long, laciniate at the apex; midlobe narrowly triangular, 2–2.5 mm long; spur straight, 7.5–16 mm long.

Habitat and distribution In forests of *Aningeria altissima* and *Morus mesozygia* and in forest partially cleared for coffee plantations between 1050 and 1200 m in Kefa. Also in Uganda, Kenya, Tanzania, Zambia, Sierra Leone and south to Zimbabwe and South Africa.

Flowering period November and December.

Conservation status Rare and possibly endangered in Ethiopia because of destruction of its forest habitat. Widespread and locally common throughout the rest of Africa.

T. bicaudata

2. T. filifolia *Schltr.*

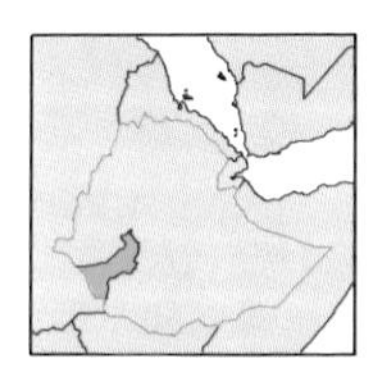
Tridactyle filiformis

The specific epithet '*filifolia*' refers to the slender, filiform leaves. Described by Rudolf Schlechter as *Angraecum filifolium* in 1905, based on his own collection from the Sanga River in Gabon. He transferred it to *Tridactyle* in 1918.

A pendent epiphyte often forming large hanging masses on trees; stems 12–80 cm × 3.5–6 mm, few-branched; roots smooth, 2–3 mm in diameter. Leaves filiform, acute, 6.5–12.5 × 0.5–1 mm, usually twisted at base to lie in one plane or less commonly conduplicate, articulated to 1–1.5 cm long leaf-sheaths. Inflorescences sessile, 1–3-flowered, 3–8 mm long; bracts amplexicaul, 0.5 mm long. Flowers dingy white, fragrant; pedicel and ovary 1–2 mm long, lepidote. Sepals lanceolate, acute, 2–3.5 × 1–1.8 mm. Petals similar. Lip 3-lobed in apical half, auriculate at the base, 1.5–3.5 × 1–2 mm; side-lobes spreading, half length of midlobe; midlobe narrowly triangular, 1 mm long; spur filiform, 6–8.5 mm long.

Habitat and distribution In forests of *Aningeria altissima* and *Morus mesozygia* and in forest partially cleared for coffee plantations at 1050 m in Kefa. Also in Uganda, Kenya, Tanzania, Zambia, Sierra Leone to Kenya and south to Zambia.

Flowering period October and November.

Conservation status Rare and possibly endangered in Ethiopia because of destruction of its forest habitat. Widespread and locally common throughout the rest of tropical Africa.

Index of scientific names

References are to main accounts.
Bold numbers indicate illustrations.

Structure of a terrestrial orchid

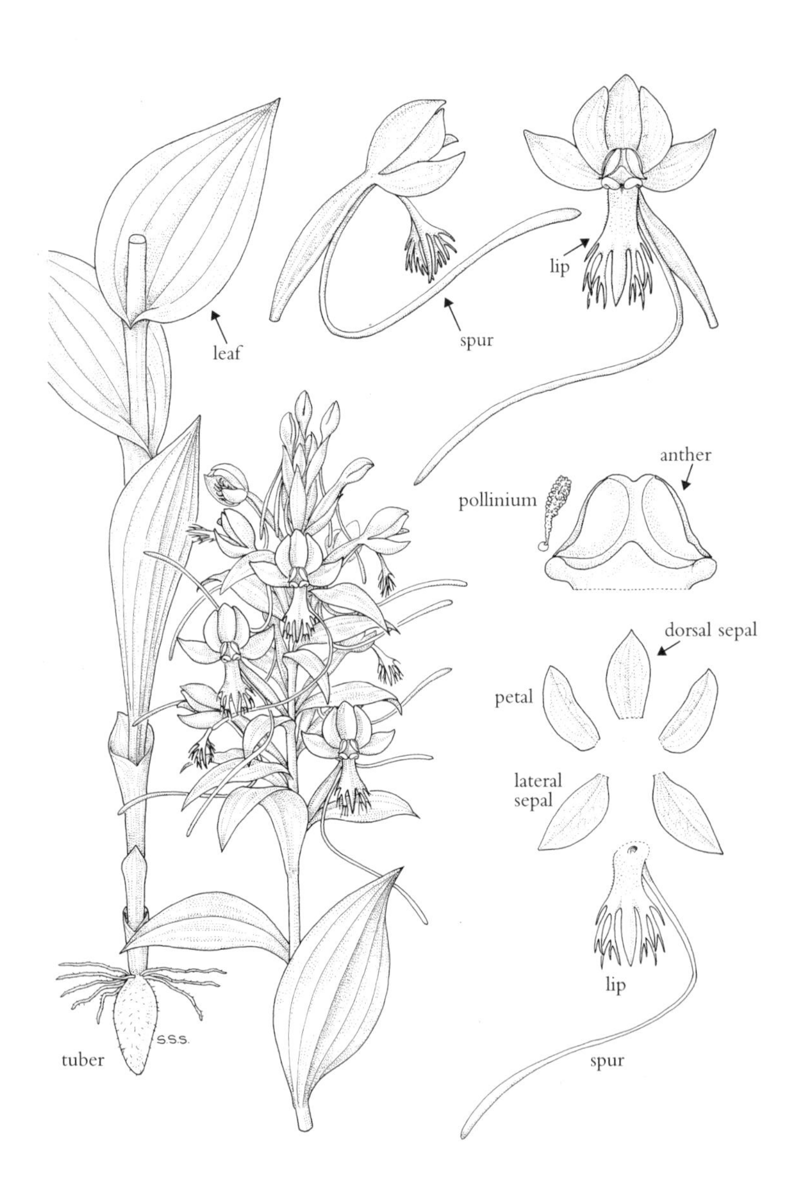

Habenaria aethiopica